Life of Fred®

Liver

Life of Fred®
Liver

Stanley F. Schmidt, Ph.D.

Polka Dot Publishing

ISBN: 978-1-937032-07-4

Library of Congress Control Number: 2012915366
Printed and bound in the United States of America

Polka Dot Publishing   Reno, Nevada

Order Life of Fred Books from:
JOY Center of Learning
http://www.LifeofFredMath.com

Questions or comments? Email the author at lifeoffred@yahoo.com

Fourth printing

Life of Fred: Liver was illustrated by the author with additional clip art furnished under license
from Nova Development Corporation, which holds the copyright to that art.

for Goodness' sake

or as J.S. Bach—who was
never noted for his plain
English—often expressed it:

Ad Majorem Dei Gloriam
(to the greater glory of God)

If you happen to spot an error that the author, the publisher, and the printer missed, please let us know with an email to: lifeoffred@yahoo.com

As a reward, we'll email back to you a list of all the corrections that readers have reported.

A Note Before We Begin

There are good ways to teach her how to cook and there are bad ways.

my daughter Jill

Rule #1: Don't just hand her a spoon, some cans of soup, and some pots and expect her to be a success.

Rule #2: Don't rush things. There is a right time to hand the cooking student a spoon, a pot, and a can of soup.

There are good ways to teach arithmetic and there are bad ways.

Rule #1: Don't just hand the student "math facts" and expect success. Most students are not robots, and they shouldn't be treated like robots.

Many traditional arithmetic books present their lessons this way:

> Here is how you do this, and here are 30 problems.
> Here is how you do this, and here are 30 problems.
> Here is how you do this, and here are 30 problems.
> Here is how you do this, and here are 30 problems.
> Here is how you do this, and here are 30 problems.

"Hey, kid. Just memorize all these facts."

Who reads those kinds of books *for fun*?
Kids will read those horrible books *just to please their parents,* not because they want to. And they end up hating math.

Arithmetic is a part of real life—such as Fred's life. If you boil it down to just a bunch of procedures, it becomes as interesting as reading a phone book.

Rule #2: Don't rush things. Learning math is not a race in which you are trying to beat others.

Brains, even those of geniuses, take time to develop. It would be silly to try and teach the antiderivative of tan x to three-month-olds.* That would be as silly as what I did over 40 years ago when I put my daughter on a bed and surrounded her with pots, cans, and a spoon.

GENERAL GUIDELINES

A) Please wait until at least the fifth grade before starting *Life of Fred: Fractions.*

Even if you were to wait until the 7th grade and have normal intelligence and drive, you would be into college calculus (*Life of Fred: Calculus*) before the end of your high school years. There is no rush.

B) Students learn algebra much, much, much, much, much better after they have started to get some hair under their arms. (That's an old saying.)

Before starting beginning algebra, check the armpits. When the student is ready, the math comes easily.

RIGHT NOW

It is the perfect time to:
- ❦ learn about livers
- ❦ practice some addition and multiplication
- ❦ experience some of Fred's adventures in life
- ❦ see a preview of fractions
- ❦ memorize the two rules of survival
- ❦ . . . and much more.

---

* . . . from calculus:

$$\int \tan x \, dx \;=\; \int \frac{\sin x}{\cos x} \, dx \;=\; \ln (\cos x) + C$$

8

HOW THIS BOOK IS ORGANIZED

Each chapter is about six pages. At the end of each chapter is a *Your Turn to Play*.

Have a paper and pencil handy before you sit down to read.

Each *Your Turn to Play* consists of about three or four questions. Write out the answers—don't just answer them orally.

After all the questions are answered, then take a peek at my answers that are given on the page following the questions.

Don't just read the questions and look at the answers.

CALCULATORS?

Not now. There will be plenty of time later after finishing *Life of Fred: Fractions* and *Life of Fred: Decimals and Percents*.

Right now, in arithmetic, our job is to learn the addition and multiplication facts by heart.

Contents

Chapter One
Toward the Camp

The bus stopped at the blood bank. Miss Ente trotted up. Fred knew that he was at the right place. He thanked the driver and hopped off the bus.

Fred was excited. Nine days of adventure at Camp Horsey-Ducky lay ahead of him. Miss Ente, the owner of the camp, was there to meet him. Fred felt very special.

Fred put down his lunch box, which he was using as a suitcase. He put his dozen set theory books and his algebra books on top of his suitcase so that they wouldn't get dirty.

The bus driver unloaded the dozen boxes that contained all the things that Fred thought he might need for camp: extra-small cowboy hat, neckerchief, rope for cows, gloves to avoid rope burns, silver spurs (with gold trim), harmonica, sundial (in case there were no clocks at the camp), mosquito spray, sun screen, campfire songbook, bow and arrows for target practice, compass, bandages, ax, canteen, poison oak soap, lantern, pancake turner (for cooking breakfast in the great outdoors), six iron frying pans of various sizes, and a case of flares (to signal for help in case of an emergency).

Fred pictured himself riding the range for nine days, probably driving cattle under the blazing sun. In the evening, he and the other cowboys and cowgirls would cook their grub over a campfire. Just before bedtime they would sit in a circle and sing "Home on the Range."

These were Fred's thoughts. Or should we say Fred's Fantasy? All he really knew was what was in that newspaper ad. It had promised life in the outdoors, horses, and thrills. Fred had never asked Miss Ente for any more information.

13

Fred had given Miss Ente the $300 camp fee when they had met at the bus stop near KITTENS. What lay ahead for Fred could be almost anything.

At one extreme, Camp Horsey-Ducky could be very tame. It might be nine days of a petting zoo. Each of the campers could walk around petting llamas, horses, cows, and ducks under shady trees. After a day or two, this would get very boring.

At the other extreme, Camp Horsey-Ducky might be a military training camp: up at 4:00 every morning, pushups, boxing, survival exercises under the blazing sun, swimming muddy rivers, and bedtime at 10:00 every night.

Fred had not asked before he signed up and gave his money.

"My oh my. You have a lot of stuff," Miss Ente said. "You won't be able to carry it all. Just put it in the camp car. Hop in and we'll get you to the camp in a jiffy."

Fred carefully loaded his twelve boxes, his math books, and his lunch box into the car. He noticed that the car had real fenders (the part over the wheels to keep mud from splashing up.)

He also noticed that his stuff had filled the car. He could not "hop in" as Miss Ente had requested. Instead, he hopped on. He rode on the hood of the car like a real cowboy.

He wondered how anyone would be able to drive.

Miss Ente threw a rope around the radiator and pulled the car down the road.

The ad for Camp Horsey-Ducky had promised life in the outdoors, horses, and thrills.

✓ It certainly was outdoors.

✓ There was a horse.

✓ Thrills? The road was very bumpy. Miss Ente was pulling the car very fast. Riding on the hood of the car was like being on a bucking horse. He was frightened that he might fall off.

He did.

He lay on the ground for a minute and wondered if he had broken anything. He hadn't. He wondered if he was bleeding. He wasn't. He wondered if he was bruised. He checked all of the 644 square inches of skin on his body. Three-fourths of his body was covered with purple bruises. The rest of his body was brown with road dirt.

$$\frac{1}{4} + \frac{1}{4} + \frac{1}{4} = \frac{3}{4}$$

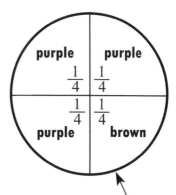

When you get to *Life of Fred: Fractions*, then we will draw pictures like this and we will add fractions like this. But not now.

Goodbye to the camp car

Fred stood up. The twelve boxes, his lunch box, his math books, the camp car, and Miss Ente were all disappearing into the distance.

small essay

Two Kinds of Hurting

Three-fourths of Fred's body was covered with purple bruises. This is one kind of hurt. Your body can hurt in many ways: cuts, bruises, toothache, broken bones, or tummy ache.

The second kind of hurt can often be worse. It is the hurt you feel in your mind: loneliness, fear, loss of a pet, or despair.

end of small essay

Fred had both kinds of hurting. Fred wept. Some of his tears rolled down his cheeks and onto his shirt. Some rolled down his nose and onto the ground.*

Fred took out his handkerchief. He decided to wipe his eyes before he blew his nose. Those two actions are not commutative. If he had blown his nose first, it would make a mess if he wiped his eyes afterward.

It's not much fun to get snot in your eyes.

---

* This is one advantage (of many) of having a large nose. Your shirt doesn't get as wet when you cry.

In doing the *Your Turn to Play*, please write out all your answers before turning to look at the answers on the next page. It is certainly easier to just look at the questions and then look at the answers, but if you do that you will not learn as much.

Your Turn to Play

1. Is addition commutative? I.e., is it always true that a + b equals b + a for any two numbers?

    If it is not true, give an example of when a + b ≠ b + a.

> Here are four common abbreviations in English:
> > i.e. = *that is to say*
> > e.g. = *for example*
> > viz. = *namely*
> > etc. = *and so on*

> ≠ means "not equal to"

2. Is subtraction commutative—i.e., is it always true that a − b equals b − a for any two numbers?

    If it is not true, give an example of when a − b ≠ b − a.

3. If Camp Horsey-Ducky were a military training camp in which you got up at 4 a.m. and went to bed at 10 p.m., how many hours would you be awake each day?

4. If you went to bed at 10 p.m. and woke at 4 a.m., how many hours would you have slept?

5. Three-fourths ($\frac{3}{4}$) of the 644 square inches of his body was bruised. How many square inches was that?

Hints: To find $\frac{2}{3}$ of 18, you multiply 18 by 2 and then divide by 3. (answer = 12)

To find $\frac{3}{5}$ of 30, you multiply 30 by 3 and then divide by 5. (answer = 18)

To find $\frac{1}{4}$ of 36, you multiply by 1 and then divide by 4. (answer = 9)

.......COMPLETE SOLUTIONS.......

1. Yes. Addition is commutative. For any two numbers it is always true that a + b = b + a.

   Multiplication is also commutative. ab = ba

2. No. Subtraction is not commutative.

   For example, $7 - 3 \neq 3 - 7$
   Neither is division.

3. 18 hours.

   4 a.m. to noon is 8 hours.
   Noon to 10 p.m. is 10 hours.

4. 6 hours.

   10 p.m. to midnight is 2 hours.
   Midnight to 4 a.m. is 4 hours.

5. 483 square inches of Fred's body were bruised.
   $\frac{3}{4}$ of 644 means 644 times 3 and then divide by 4.

```
     644              483
   ×   3         4)1932
   ------          16
    1932           --
                   33
                   32
                   --
                   12
                   12
```

Chapter Two
Survival

Fred looked down the road. How far was it to Camp Horsey-Ducky? It might be a mile or it might be 50 miles. He might have to walk for days to get there. Would he have to spend the night sleeping beside the road? Would he be eaten by lions?*

Last year Fred had read Prof. Eldwood's *Modern Guide to Survival*. He tried to remember what that book said. It told nothing about how to land a plane, but this was not surprising. The book was written in 1870, and airplanes were invented in the early 1900s.

It said nothing about surviving if your parachute fails to open. There wasn't much need for parachutes until airplanes were invented.

Fred remembered Eldwood's First Rule of Survival:

> ## Don't PANIC.
> It makes you do stupid things.

His Second Rule of Survival: If you do stupid things, you might end up dead.

Fred put his fingers on his wrist and took his pulse. 160 beats/minute. That was not a good sign. He sat down beside the roadway and tried to calm himself.

---

* Fred was not thinking very clearly. The chances of being eaten by lions if you are in Kansas are very small, because the only lions in Kansas are in zoos. Zoos have very strict rules about not climbing into lion cages.

He thought When did I get off the bus? It was 4:10. I remember seeing that on the clock that was on the Kansas Blood Bank building.

How long did it take me to pack the camp car? About four minutes.

How long did I ride on the hood of the car before I fell off? About a minute.

Fred's pulse had slowed to 90 beats/minute. He was getting less panicked and less stupid.

Instead of looking at a possible 50-mile hike down the road to Camp Horsey-Ducky, another possibility occurred to him. He slowly turned around and saw

The clock now read a quarter after four. Fred's pulse dropped to 60. He was feeling much better.

He stood up and walked back to the building. Prof. Eldwood was right: panicking doesn't help.

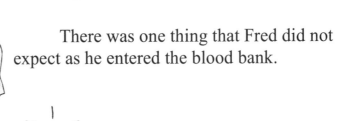

There was one thing that Fred did not expect as he entered the blood bank.

He let out a little scream. For once, he didn't faint. His pulse was back up to 160.

Don't be afraid. That lion is just a big doll that the blood bank owns. It is not meant to scare little boys. They put it here to say that blood donors have the courage of a lion.

Dr. Morningstar

It was Dr. Morningstar, the doctor who had examined Fred for his camp physical.

Fred asked, "Dr. Morningstar, what are you doing here? I just saw you a couple of hours ago at KITTENS Hospital."

"Hi, Freddie. I work from 7 a.m. until 3 p.m. at the hospital. I'm here as a volunteer to make a blood donation."

Just then, her name was called, and she headed into the blood donation area.

Fred headed over to the lion doll and gave it a pat on its paw. Once he understood, he wasn't frightened anymore.*

"Do you like our lion?" the receptionist asked.

Fred smiled and said, "Yes, I do. Where did you get such a big doll? He is twice as tall as I am."

"One of our donors, Mr. Clef, made it for us. He teaches music at KITTENS University, and he also makes dolls as a hobby. Someday when you get older, you might be a student at KITTENS."

Fred giggled to himself. He wanted to tell the receptionist that he would probably never be student there, but he decided to forego all the explanation that it would involve. Instead, he asked, "By the way, when does the next bus come that is headed back to KITTENS?"

---

* That is true of many things in life.

"I'm afraid that the last bus of the day came here about ten minutes ago. Wasn't that your mommy that went in to donate a moment ago?"

"No. That was Dr. Morningstar, my doctor. I came on the 4:10 bus. I'm afraid that I am stranded."

The receptionist didn't quite understand. She asked, "Is there someone at your house that we could call?"

"The only one there is Kingie, but he doesn't have any legs. I don't have any money with me right now since I lost my lunch box, which had my money in it."

"Don't you worry. I'll be back in a minute. Just have a seat. Here's a piece of paper and a pencil to draw with." The receptionist headed into the back office.

In the back office, the receptionist explained to her coworkers that there was this little three-year-old boy out in the reception area who got off the bus by accident, who lives near KITTENS, who has a father who lost both of his legs either in a war or because of diabetes, and who has lost his lunch box. (The only thing she was right about was Fred's losing his lunch box.) She asked them, "What shall we do? We can't just kick him out when the blood bank closes."

Minnie said, "My shift is over at 5. That's only about a half hour from now. I'll drive him home. I live near KITTENS."

The receptionist came out and told Fred, "Minnie is one of my coworkers. She said she would be glad to take you home after her shift is over at five."

Time Out!

English! Nobody really understood what was being said. The receptionist thought that

Fred had gotten off the bus by accident. He had *deliberately* gotten off when he saw Miss Ente.

She thought that Kingie was Fred's father.

When the receptionist said that Minnie "would be glad to take you home," Fred thought that Minnie would be taking him to *her* house.

Your Turn to Play

1. If the lion was twice as tall as Fred, how tall was the lion? (Fred is three feet tall.)

2. If Fred's pulse had been 160 beats/minute and had slowed to 90, how much was the decrease?

3. Three-fifths ($\frac{3}{5}$) of the dolls that Mr. Clef makes are lion dolls. If he made 75 dolls in May, how many of them were lion dolls?

4. Dr. Morningstar works from 7 a.m. to 3 p.m. How many hours is that?

5. There are things that are true that maybe shouldn't be said. The two rules of survival that Prof. Eldwood wrote are true:

#1 Don't panic. It makes you do stupid things.

#2 If you do stupid things, you might end up dead.

Here is a tricky question: What would be bad about telling people his Second Rule of Survival?

(Please think about this for a moment. Don't just turn the page and look at my answer.)

```
.......COMPLETE SOLUTIONS.......
```

1. If the lion doll is twice as tall as three-foot Fred, it would be six feet tall.

2. It had decreased by 70 beats/minute.

3. Mr. Clef had made 45 lion dolls.

To find $\frac{3}{5}$ of 75, you multiply 75 by 3 and then divide by 5.

$$
\begin{array}{r}
75 \\
\times\ 3 \\
\hline
225
\end{array}
\qquad
\begin{array}{r}
45 \\
5\overline{)225} \\
\underline{20} \\
25 \\
\underline{25}
\end{array}
$$

4. 8 hours.

   From 7 a.m. to noon is 5 hours.

   From noon to 3 p.m. is 3 hours.

5. Prof. Eldwood was not thinking very clearly when he wrote the Second Rule of Survival: " . . . you might end up dead."

   That would cause some people to panic. And then they would be violating the First Rule of Survival: "Don't PANIC."

small essay
Taking Exams Like the SAT

There are three mental states for you to choose among:
State #1: You PANIC. You get all steamed up. Your brain freezes. You can hardly hold your pencil.
State #2: You are alert. You had a good night's sleep.
State #3: Your mind wanders. You are thinking about a half dozen things at the same time rather than concentrating. This is called multi-tasking. My grandmother used to call this being scatterbrained.

end of small essay

Chapter Three
A Ride Home

F red couldn't believe his ears when he heard the receptionist say that Minnie was going to take him home. He thought she meant that Minnie was going to take him to her house and keep him.

He wondered Can blood bank workers do this? Can they collect any stranded kids that they find? Does Minnie have a whole houseful of kids? Can't I just go back to my office in the Math Building?

"Hi. I'm Minnie. The receptionist told me that you could use a ride back to your house."

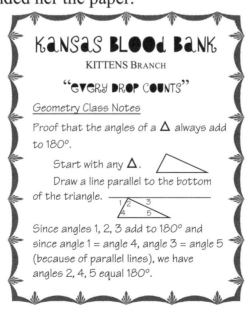

Minnie's English was a lot better than the receptionist's. When she said, ". . . *back* to *your* house," Fred's fears melted away.

"It looks like you were doing some drawing on the paper that the receptionist gave you. May I see it?"

Fred handed her the paper.

KANSAS BLOOD BANK
KITTENS BRANCH

"EVERY DROP COUNTS"

Geometry Class Notes

Proof that the angles of a △ always add to 180°.

    Start with any △.

    Draw a line parallel to the bottom of the triangle.

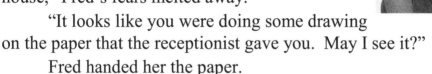

Since angles 1, 2, 3 add to 180° and since angle 1 = angle 4, angle 3 = angle 5 (because of parallel lines), we have angles 2, 4, 5 equal 180°.

Minnie was expecting to see some kid's drawings. When she saw Fred's <u>Geometry Class Notes</u>, she was surprised and said, "You are so young. How could you be studying geometry?"

"This is just high school geometry," Fred explained. "I learned this about five years ago.* I guess I labeled the page incorrectly. I should have written <u>Geometry Lecture Notes</u>. I teach geometry daily at 11 o'clock at KITTENS."

"Oh, you silly boy! No three-year-old teaches at a university," she said.

"I'm not three. I'm five. I'm just short for my height—I mean for my age. My class is in the giant lecture hall in the Archimedes Building. You are welcome to come and see for yourself."

"Okay," she said. "Tomorrow is my day off. I'll be there."

Fred turned a little pink and said, "I just remembered. There won't be any class tomorrow. The university president has canceled all the classes."

"I knew it! You little lamb. I knew you were just fooling with me."

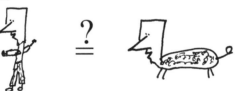

Fred had never been called a little lamb. He knew that Minnie didn't mean that literally. No wool. No four legs. *You are a little lamb* is a metaphor. It meant that Fred was cute and cuddly.

---

* It was a little more than five years ago. Details are in the story of Fred's earliest years as told in *Life of Fred: Calculus*.

Intermission

Some readers might be curious how Fred could write that angle 1 = angle 4 in his geometry notes.

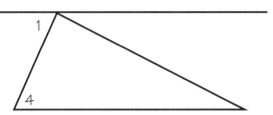

Fred was using a statement that he had proved in the previous class meeting. (Statements that are proved in math are called theorems.)

Theorem: If two lines are parallel, then the alternate interior angles are equal.

All this will be explained when you get to geometry.

"I'll go get my purse," Minnie said, "and we can get going. Did you bring anything with you on your bus trip?"

Fred answered truthfully (although it would have been easier to lie), "I brought my lunch box with my clothes in it, a dozen set theory math books, and twelve boxes that contained my extra-small cowboy hat, neckerchief, rope for cows, gloves to avoid rope burns, silver spurs (with gold trim), harmonica, sundial (in case there were no clocks at the camp), mosquito spray, sun screen, campfire songbook, bow and arrows for target practice, compass, bandages, ax, canteen, poison oak soap, lantern, pancake turner (for cooking breakfast in the great outdoors), six iron frying pans of various sizes, and a case of flares (to signal for help in case of emergency)."

"Oh you big kidder," Minnie laughed. "And where is all this stuff?"

Fred gulped. He was afraid she wouldn't believe him when he told her that it was in the 1927 camp car that Miss Ente was pulling down the roadway. But she had asked, so Fred told her.

"You are such a card!*" she said. "The next thing you are going to tell me is that Miss Ente is a jet plane that towed the car."

Fred giggled. "You are so silly, Minnie. Miss Ente isn't a jet plane. She is a talking horse that owns Camp Horsey-Ducky."

Minnie took Fred's hand and they walked out to her car.

The car was parked at the edge of the parking lot. Minnie explained that the car had a little problem with leaking oil, so she always parked it away from the other cars.

Minnie opened the trunk and took out a quart of oil. She had several cases of oil in the trunk. She walked around to the front of the car, opened the hood, and poured the oil into the correct hole.

She told Fred that she didn't know a lot about cars. When she first owned this car, she accidentally put a quart of oil into the wrong hole. It was the hole for window washing fluid. Every time she used the window washers she had to go to a car wash to get the oil off her windshield.

---

* The 13th definition of a *card* in my dictionary is "a witty and amusing person."

Your Turn to Play

1. Minnie had many cases of oil in her trunk. She needed a lot of oil. A case of Oily Oil™ was a box that was 4 cans wide, 3 cans deep, and 2 cans tall.

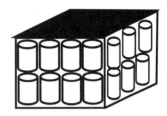

How many cans are in a case of Oily Oil™?

2. Oily Oil™ cost 79¢ per can when bought in a case. How much would a case cost?

3. Minnie explained to Fred that she would put a quart of oil in her car every morning before driving to work. She would put a quart in when she parked the car in the blood bank parking lot. She would add a quart before and after lunch, a quart before driving home, and a quart when she got home. Six quarts each day. How many days would it take Minnie to go through a case of oil?

4. Minnie said that she had purchased the car two months ago from a "very nice man" at Coalback Used Cars. She was happy to get this car because it was the last car on the lot. The "very nice man" had told Minnie that his cars were so popular she was lucky he had any cars left. (In truth, this oil-leaking car was the only one that Coalback Used Cars had ever had on its lot. Coalback had been trying to sell it for weeks.)

He even gave her a lifetime guarantee. Read it and guess why the guarantee was worthless.

Lifetime Iron-Clad Guarantee

Coalback Used Cars unconditionally guarantees that the car you have just purchased will be free of defects of any kind. Any defect will be repaired at no cost to you by Coalback Used Cars. Your complete satisfaction is assured. This guarantee is good for a lifetime: as long as you, the car, and Coalback Used Cars exist.

```
....... COMPLETE SOLUTIONS .......
```

1. $4 \times 3 \times 2 = 24$ cans in a case

2. We want 24 cans @ 79¢ per can.   (@ means *at*)
If you don't know whether to add, subtract, multiply, or divide, the General Rule is to restate the problem using simple numbers. If we wanted 3 cans @ 2¢ per can, it would cost 6¢. We multiplied.

$$\begin{array}{r} 79 \\ \times\,24 \\ \hline 316 \\ 158 \\ \hline 1896 \end{array}$$

A case would cost 1,896¢
or $18.96.

3. Minnie uses 6 quarts per day. We want to know how long it would take to go through a case, which is 24 quarts.

   Using the General Rule, we restate the problem using simple numbers. If she was using 2 quarts a day and a case was 6 quarts, it would take her 3 days. We divided.

$$6\overline{)24} = 4$$ It would take her 4 days.

4. The last line of the guarantee reads, "This guarantee is good . . . as long as . . . Coalback Used Cars exists." The minute that Minnie bought the one-and-only car that Coalback Used Cars had on its lot, it went out of business.

For much of your adult life, you will be asked to sign applications, contracts, mortgages, and leases. And they will often contain pages of fine print.

   If you sign and things end up in court, you are obligated by the fine print, whether you read it or not.

   A lease might say in the 14th paragraph, if you don't pay your rent, the landlord can take your car and your dog.

Chapter Four
Car Repair

As Minnie drove out of the parking lot and onto the highway, Fred noticed a line of oil on the road. It was easy to tell which way Minnie drove each morning and evening.

Minnie's commute each morning and evening

Fred thought that spending $18.96 every four days to buy oil was getting a little expensive.[*]

He didn't know much about cars, but he did know that sometimes you can save money if you buy in larger quantities. Minnie had been buying cases of oil, and that was cheaper than buying single quarts of oil. Fred suggested that she might consider buying a drum of oil and then run a hose from the drum to underneath the hood. Then she would save both time and money. She wouldn't have to keep adding quarts of oil six times a day.

"Oh! Why didn't I think of that?" Minnie squealed. "You are such a smart little monkey."

Fred had been a *little lamb*, and was now a *smart little monkey*. Minnie liked metaphors.

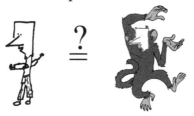

---

[*] Expensive? Some people spend more than that every four days on cigarettes.

Minnie spotted a car repair place. She was going to take Mr. Smart Monkey's suggestion. Many car repair places close around 5 p.m. That is not very convenient for people who work from 8 to 5.

Edward's Afternoon Auto was different. They opened in the afternoon and stayed open until 9 p.m. Because of that, they had a lot of business.

small essay

How to Be a Super Success

If you own a business, there is only one rule. That rule will bring you success beyond your dreams.

The Rule: Offer your customers what THEY want or need.

It is that simple. It is not what you think is cool. It is what they think is cool. You don't try to sell cigars to three-year-old girls when what they want are Rag-A-Fluffy dolls.

Work like crazy to deliver:

☆ the product they want

☆ good quality

☆ good price

| Customers also want these two things. |

and you will be a winner.

They will be happy, and you will be successful.

end of small essay

Minnie drove into the parking lot of Edward's Afternoon Auto. As they got out of the car, Edward came over to greet them. They didn't have to go into some dirty office that smelled like tires and wait in line to be helped.

"Good afternoon. I'm Edward. How may I help you?"

Minnie pointed to the puddle of oil that was forming. Before she could mention Fred's idea of installing a drum of oil on the trunk of the car, Edward smiled and said, "It looks like we've got a bit of a leak. We can get to it right now. It should take less than an hour."

Minnie said, "I'm famished. Is that place next door any good?" She pointed to *K's Afternoon Dining*.

"It's almost too good," Edward said. "It opened a week after I opened my shop here. It has the same hours as my shop. The owner, Katherine, serves up some tasty food. I've put on a couple of pounds eating there."

Minnie took Fred's hand and they headed over to *K's*. She didn't ask if Fred was hungry. He looked like he hadn't eaten in weeks.

They headed inside and sat down. The sign on the wall announced: ❁ *Great food!*

    ❁ *Low prices!*

    ❁ *Large menu!*

Katherine must have read that famous essay about "How to Be a Super Success." She offered the product customers want, good quality, and good price.

The waitress gave Fred the child's menu and some crayons. Fred had a choice of a hot dog on a stick or macaroni and cheese.

Minnie got the large menu. It was huge. It was like a book. At *K's*, when they advertised "large menu" they really meant it.

The menu was divided into chapters.

The first chapter was food from Armenia. The second chapter was food from Norway. The third chapter was food from Guam. Minnie turned to the index at the back of the menu. (Very few menus are so large that they need an index.) The index read: *food from* . . . Afghanistan, Akrotiri, Albania, Algeria, American Samoa, Andorra, Angola, Anguilla, Antarctica, Antigua and Barbuda, Argentina, Armenia, Aruba, Ashmore and Cartier Islands, Australia, Austria, Azerbaijan, (The) Bahamas, Bahrain, Bangladesh, Barbados, Bassas da India, Belarus, Belgium, Belize, Benin, Bermuda, Bhutan, Bolivia, Bosnia and Herzegovina, Botswana, Bouvet Island, Brazil, British Indian Ocean Territory, British Virgin Islands, Brunei, Bulgaria, Burkina Faso, Burma, Burundi, Cambodia, Cameroon, Canada, Cape Verde, Central African Republic, Chad, Chile, China, Christmas Island, Clipperton Island, Cocos (Keeling) Islands, Colombia, Comoros, Congo, Democratic Republic of the, Congo, Republic of the, Cook Islands, Coral Sea Islands, Costa Rica, Cote d'Ivoire, Croatia, Cuba, Cyprus, Czech Republic, Denmark, Dhekelia, Djibouti, Dominica, Dominican Republic, Ecuador, Egypt, El Salvador, Equatorial Guinea, Eritrea, Estonia, Ethiopia, Europa Island, Falkland Islands (Islas Malvinas), Faroe Islands, Fiji, Finland, France, French Guiana, French Polynesia, French Southern and Antarctic Lands, Gabon, Gambia, Georgia, Germany, Ghana, Gibraltar, Glorioso Islands, Greece, Greenland, Grenada, Guadeloupe, Guam, Guatemala, Guernsey, Guinea, Guinea-Bissau, Guyana, Haiti, Heard Island and McDonald Islands, Holy See (Vatican City), Honduras, Hong Kong, Hungary, Iceland, India, Indonesia, Iran, Iraq, Ireland, Isle of Man, Israel, Italy, Jamaica, Jan Mayen, Japan, Jersey, Jordan, Juan de Nova Island, Kazakhstan, Kenya, Kiribati, Korea, North, Korea, South, Kuwait, Kyrgyzstan, Laos, Latvia, Lebanon, Liberia, Libya, Liechtenstein, Lithuania, Luxembourg, Macau, Macedonia, Madagascar, Malawi, Malaysia, Maldives, Mali, Malta, Marshall Islands, Martinique, Mauritania, Mauritius, Mayotte, Mexico, Micronesia, Federated States of, Moldova, Monaco, Mongolia, Montserrat, Morocco, Mozambique, Namibia, Nauru, Navassa Island, Nepal, Netherlands, Netherlands Antilles, New Caledonia, New Zealand, Nicaragua, Niger, Nigeria, Niue, Norfolk Island, Northern Mariana Islands, Norway, Oman, Pakistan, Palau, Panama, Papua New Guinea, Paracel Islands, Paraguay, Peru, Philippines, Pitcairn Islands, Poland, Portugal, Puerto Rico, Qatar, Reunion, Romania, Russia, Rwanda, Saint Helena, Saint Kitts and Nevis, Saint Lucia, Saint Pierre and Miquelon, Saint Vincent and the Grenadines, Samoa, San Marino, Sao Tome and Principe, Saudi Arabia, Senegal, Serbia and Montenegro, Seychelles, Sierra Leone, Singapore, Slovakia, Slovenia, Solomon Islands, Somalia, South Africa, South Georgia and the South Sandwich Islands, Spain, Spratly Islands, Sri Lanka, Sudan, Suriname, Svalbard, Swaziland, Sweden, Switzerland, Syria, Taiwan, Tajikistan, Tanzania, Thailand, Timor-Leste, Togo, Tokelau, Tonga, Trinidad and Tobago, Tromelin Island, Tunisia, Turkey, Turkmenistan, Turks and Caicos Islands, Tuvalu, Uganda, Ukraine, United Arab Emirates, United Kingdom, United States, Uruguay, Uzbekistan, Vanuatu, Venezuela, Vietnam, Virgin Islands, Wake Island, Wallis and Futuna, West Bank, Western Sahara, Yemen, Zambia.

Minnie was overwhelmed. She turned back to the first chapter and looked at the Armenian food.

First were the Armenian soups: Arganak Blghourapour, Bozbash, Brndzapour, Dzavarapour, Flol, Harissa, Katnapour, Katnov, Kololik, Krchik, Mantapour, Matsnaprtosh, Putuk, Sarnapour, Snkapur, Tarkhana, Vospapour, and Pekhapour.

Then the Armenian main courses: Fasulya, Ghapama, Kchuch, Tjvjik, Satsivi, Basturma, Yershig, and Kiufta.

She skipped over the fish dishes, the breads, and the sweets (including Kadaif, which is dough with cream and chopped walnuts in a sugar syrup).

The waitress came back. Minnie didn't know what to do. She picked an item at random and pointed to the main course Tjvjik. She couldn't pronounce it (and she didn't know what she was ordering).

Fred asked for a glass of water. He said that he hadn't decided yet.

There are very few things in mathematics that every educated person is expected to know. When you get to algebra and geometry, you will learn the Pythagorean theorem, which says that in any right triangle, $a^2 + b^2 = c^2$.

Many adults don't know that.

But every educated person should know the addition and multiplication tables. $7 + 8 = 15$ and $7 \times 8 = 56$.

This is the time in your life to learn them.

Please do not use a calculator in this *Your Turn to Play*.

Your Turn to Play

1. A drum of motor oil is 55 gallons. A gallon (of anything) is 4 quarts. How many quarts of oil are in a drum of oil?

2. Katherine has been starting restaurants for years. **K's Afternoon Dining** is the third one she has started this year. She earns about $400,000 each year before taxes. The federal income tax, state income tax, property taxes, licenses, and payroll taxes take two-fifths ($\frac{2}{5}$) of her income. How much does she pay each year?

3. Suppose a new medium-priced car cost $32,000. How many of those cars could Katherine buy each year with the money that all the layers of government take from her in taxes?

4. A case of oil cost $18.96. Minnie had been spending 1,896¢ every four days for oil. How much is that per day? (We will work in cents so that you won't have to deal with decimals.)

·······COMPLETE SOLUTIONS·······

1. We want to change 55 gallons into quarts. There are 4 quarts in a gallon. Do we add, subtract, multiply, or divide? The General Rule is to restate the problem using simpler numbers and notice which operation (+, −, ×, or ÷) you used.

If we had 2 gallons (instead of 55) and we wanted to change it into quarts, that would be 8 quarts. We multiplied.

So we multiply 55 by 4.

$$\begin{array}{r} 55 \\ \times\ 4 \\ \hline 220 \end{array}$$

There are 220 quarts in a 55-gallon drum of oil.

2. To find $\frac{2}{5}$ of $400,000, you multiply 400,000 by 2 and then divide by 5.

$$\begin{array}{r} 400,000 \\ \times\qquad 2 \\ \hline 800,000 \end{array}$$

$$\begin{array}{r} 160000 \\ 5\overline{)800000} \\ \underline{5} \\ 30 \\ \underline{30} \end{array}$$

The government takes $160,000 from Katherine each year.

3. How many $32,000 cars would $160,000 buy? Restating, using easier numbers: How many $2 cars would $8 buy? It would be 4 of them. We divided.

$$\begin{array}{r} 5 \\ 32000\overline{)160000} \\ \underline{160000} \end{array}$$

4. One-fourth ($\frac{1}{4}$) of 1,896¢.

$$\begin{array}{r} 474 \\ 4\overline{)1896} \\ \underline{16} \\ 29 \\ \underline{28} \\ 16 \\ \underline{16} \end{array}$$

Minnie spends 474¢ (= $4.74) each day.

Chapter Five
Ordering Tjvjik

The waitress brought Fred his glass of water and Minnie her Tjvjik. If she had ordered the Ghapama from the Armenian chapter of the menu, she would have received a pumpkin stew. If she had ordered the Satsivi, she would have received pieces of roast chicken in a walnut sauce.

Tjvjik is a dish of fried beef liver and kidneys served with onions. Minnie was delighted.

"My mom used to make liver and onions every Wednesday night," Minnie said. "All of our family loved it, but many people don't like the taste. Everyone likes different foods. I once met someone who didn't like chocolate ice cream."

At this point, Minnie started eating and Fred took a sip of his water.

After Minnie had eaten for a while, she paused and said, "I have fond memories of going to the grocery store with my mother when I was a kid. When we got to the meat section, she would get steak for Sunday nights and hamburger for Tuesday. I remember her looking at all the packages of liver for Wednesday's dinner. The meat was a dark reddish brown.

"Humans have livers also. Your liver is like a roof over your stomach." She pointed to Fred's tummy.

"My husband is a hepatologist.* He never stops talking about how wonderful livers are. When some people doodle,

---

* Hepatologist = one who studies livers. When you see the prefix *hepat* or *hepato*, you can guess that it has something to do with livers. Hepatic = liver-colored. Hepatitis = inflamed liver. Hepatocyte = a liver cell.

they draw pictures of hearts or of horses. He draws pictures of livers."

    When Fred doodled, he sometimes drew geometry diagrams.

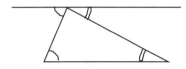

He sometimes scribbled equations.

$$2x - 7 = 11$$
$$2x \quad\ = 18 \text{ Add 7 to both sides.}$$
$$x \quad\ = 9 \qquad \text{Divide both sides by 2.}$$

He sometimes drew silly pictures.

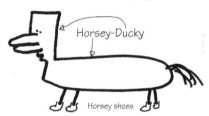

    Minnie continued, "When I first met him, he asked me to name the largest organ inside the human body. I told him I didn't know. He asked me to guess. I looked at his name tag,

which read | Karl hepatologist | and so I guessed that the liver was

the largest organ inside the human body. I think that was the moment that Karl really started to like me.

    "He told me that the liver was larger in volume than all of the rest of the organs of the body put together. He said that one-fifth of the blood in my body was inside the liver."*

---

* Actually, somewhere between one-fifth and one-fourth of the blood in the body is inside the liver.

Fred drew a picture:

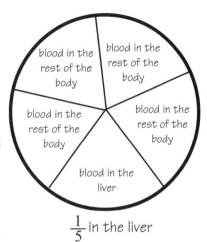

$\frac{1}{5}$ in the liver

Fred asked, "What is all that blood doing inside the liver?"

After being married to a hepatologist for five years, Minnie knew the answer to most questions about livers. "As your blood passes through your liver, it gets changed in hundreds of different ways. The liver works

☆ with the fats in your blood

  ☆ with the proteins

    ☆ with the carbohydrates

      ☆ with the sugars

        ☆ with the amino acids

          ☆ with the iron.

"It makes blood proteins and inserts them into your blood stream at just the right rate. If it did it too quickly, your blood would get too thick."

Fred mentally finished her thought And if my liver did it too slowly, my blood would get too thin. (Fred was right.)

Minnie said, "Karl once spent an evening explaining to me how the liver changes some of the bacon and sausage that we had for breakfast (saturated fats) into cholesterol."

Fred asked, "I thought cholesterol was supposed to be bad for you. Why does the liver make it?"

"Heavens, no," Minnie said. "Your body needs cholesterol. It is only when there is too much—or too little—that things get bad."

Karl had mentioned hypercholesterolemia (HI-per-ka-LES-tera-LEE-me-eh) to Minnie, but she was afraid to say a nine-syllable word in front of a five-year-old. The word means too much cholesterol in the blood.

Nor did Minnie want to mention hypocholesterolemia, which is another nine-syllable word meaning abnormally low levels of cholesterol in the blood.

hyper = too much

hypo = too little

Fred would have loved to have learned those big words. Some little kids might have nightmares with giant words like *hypercholesterolemia* chasing them down the hallway. Fred was different. Big words were like giant teddy bears that he could snuggle with at night.

As every English major knows, words have power.*

_____

* Several hundred years ago (actually, in 1961) when I, your author, entered college, they administered a test to all incoming freshmen. The test was used to determine where you would start in mathematics and in English.

Part of the English exam was vocabulary. Now, over fifty years later, I still remember two words from that test: *saturnalia* and *saturnine*. I didn't know what either of them meant.

When I got home that day, I looked them up in my dictionary.

Saturnalia = wild partying.

Saturnine = gloomy and lacking in energy. Those words seemed opposite in meaning until I realized that after saturnalia you feel saturnine.

Reading good books helps build your vocabulary. Watching television does not help much at all. When is the last time you heard a word on television that was new to you?

Because of the previous page in this book, your vocabulary might now be four words larger.

Your Turn to Play

1. Two pages ago, Fred drew a **pie chart** to illustrate how much blood was in the liver compared with the blood in the rest of the body. That chart showed that one-fifth ($\frac{1}{5}$) of all your blood is in the liver.

We know that somewhere between one-fifth and one-fourth of your blood is in your liver.

Draw a pie chart illustrating that one-fourth ($\frac{1}{4}$) of your blood is in your liver.

2. Find the **product** of 67 and 89. (*Product* means multiply.)

3. Find the **sum** of 67 and 89. (*Sum* means add.)

4. (Question for artists) Fred drew a picture of what he thought a horsey-ducky would look like: the body of a horse and the head of a duck.

Draw a picture of how Fred might have drawn a ducky-horsey.

·······COMPLETE SOLUTIONS·······

1.
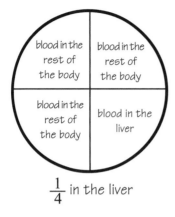

$\frac{1}{4}$ in the liver

It is okay if you drew your pie chart this way.

2.
```
    89
  × 67
  ─────
   623
   534
  ─────
  5963
```
$89 \times 67 = 5{,}963$

3.
```
    89
  + 67
  ─────
   156
```
$89 + 67 = 156$

4. Here is how Fred drew a ducky-horsey. You can probably draw better than he can.

Horsey and Ducky are not commutative. $\neq$

horsey-ducky ≠ ducky-horsey.

Chapter Six
A Plug

Minnie had finished her Tjvjik. Fred had only taken one sip of water. He was full. Before she told him about how big livers are and what they do, Fred had heard only one thing about liver: cirrhosis (sih-ROW-sis) of the liver. But he didn't know what that meant.

So he asked.

"It's a way to die," Minnie explained. "Cirrhosis of the liver is the 12th most popular way to die of disease in this country."

Fred winced* at the word *popular*. He would have said that cirrhosis is the 12th most common way.

Minnie said, "Cirrhosis is scarring of the liver. You know about scars on the skin. The liver can also get scars."

"People cut their livers?" Fred asked.

"No. That is not the way they damage their livers. The most common way is by drinking a lot of alcohol. One of the jobs of your liver is to purify your blood. Each time the liver cleans alcohol out of the blood stream, some cells of the liver die and turn into scar tissue. Not a lot, just some.

"Your liver can take quite a bit of abuse, but heavy drinking over many years and you have . . ."

Fred mentally completed her sentence . . . the 12th most popular way to die of disease.

Before Minnie had explained to him what cirrhosis was, Fred had thought that the worst aspect of a heavy drinking session was that it made you stupid.

---

* Winced = to respond involuntary to pain or disgust. Your body tenses or you make a face as if you had just sat on a tack.

They headed back to Edward's Afternoon Auto. When they walked in, Edward was laughing.

"Lady, this was the easiest car repair I've done in years." He handed her the keys. "When they change your oil in a shop, there are three steps. First, they remove the drain plug so that the old, dirty oil can pour out. Then they replace the plug and fill the car with new oil. Whoever changed the oil forgot to put the plug back in. Any oil you poured in would just drain out."

Minnie asked, "How much do I owe you?"

"Nothing. The plug cost two bits,* but the laugh is payment enough." Minnie put five quarts of oil in and gave Edward the rest of the oil she had in the trunk as a thank-you gift.**

"I guess it's time we get you back home. I bet your father is worried about you."

Father? Fred thought. Then Fred realized that when he had mentioned Kingie, they thought he was talking about his father. Fred told Minnie who Kingie really was.

He also told her where he lived. She pulled out a map while she was driving. The radio was on. She was talking with Fred. Her cell phone rang.

And she sailed r i g h t through a red light.

One of the drawbacks of multi-tasking.

Fred quickly put on every seatbelt he could find.

---

* Two bits is old-fashioned talk for 25¢.

** Most people do not have to add oil to their cars between oil changes.

When Minnie ran the red light, there were three possibilities:

(1) Nothing might have happened. This would be the best outcome. However, nothing didn't happen.

(2) A policeman might have seen her run the red light, pulled her over, and given her a ticket. That would have cost Minnie several hundred dollars. That didn't happen.

(3) An accident. Minnie never saw what hit her. She was too busy reading the map, listening to the radio, talking with Fred, and answering her cell phone.

There are three possible vehicles that might have hit Minnie's car:

(1) A motorcycle. This would have caused little damage to Minnie's car and a lot to the motorcycle.

(2) Another car like Minnie's. Both would have been equally damaged.

(3) A big truck. When a truck and a car collide, the car usually receives the most damage.

Here is a hint as to what happened.

Accident Report

```
1.   Fred was okay.
2.   Minnie bumped her head pretty hard.
She was knocked out.
```

3. The car was a complete wreck. Only
two parts of the car survived: the radio
was still playing and the oil drain plug
had not fallen out.

The truck driver called 9-1-1 and reported that the driver
of the car was unconscious. The 9-1-1 dispatcher said that help
was on the way.

In a couple of minutes, the police, the paramedics from
the fire department, and an ambulance arrived.

Minnie didn't have her seatbelt on. She had taken it off
"just for a moment" when she had reached over to pull a map
out of the glove box.

It seemed like chaos* to Fred, but everyone was doing
the job they were trained to do. The policeman was
interviewing the truck driver to find out what had happened.
The paramedics helped move Minnie to the ambulance. A tow
truck from Edward's Afternoon Auto towed Minnie's car away.

One of the paramedics came to comfort Fred. He said,
"Your mom just bumped her head. She'll be okay." He didn't
want to scare this little boy by mentioning that she would
probably need X-rays to find out if any bones were broken.
The paramedic figured that Fred was about three years old and
might not understand what X-rays are.

Fred said, "I think it would be too early to rule out an
extradural hemorrhage (HEM-more-ridge). That condition will
occur, as you know, in about 2% of those admitted to a hospital
as a result of a blow to the head."

-----

* KAY-os Chaos is a state of disorderly confusion.
  The word *chaos* and *yacht* are great words to use in a spelling contest.

The paramedic couldn't believe what he was hearing. Three-year-olds don't use that language.

"Or the possibility," Fred continued, "of a subdural or a subarachnoid hemorrhage also exists."

Translation
for those readers who have not read
as much medical literature as Fred has

Most people know what a bruise is. A blow can cause blood vessels to break, and you see a purple mark under your skin.

Brains are covered by three thin layers with blood vessels in each of them. A severe head injury can break those blood vessels. Depending on which layer the broken blood vessels are in, you get either an extradural, a subdural, or a subarachnoid hemorrhage. In baby talk, it is called a "brain bruise."

Your Turn to Play

1. You wear a helmet when you ride a bicycle. Most bicycle riders cannot name the three kinds of hemorrhages that wearing a helmet can help prevent. You can. Name them.

2. If you had to draw a hundred stars, you could make them into a rectangle that was 5 rows by 20 columns:

Could those 100 stars also be arranged into a square?

3. (Question for artists) Two percent (2%) means 2 out of 100. Illustrate this by drawing 100 circles and coloring in two of them.

4. Find $\frac{6}{7}$ of 413. (If you need a reminder of how to do this, look back to the bottom of page 17.)

.......COMPLETE SOLUTIONS.......

1. The three kinds of brain hemorrhages are extradural, subdural, and subarachnoid.

2. Yes, 100 stars can be arranged in a square. It will have 10 rows and 10 columns.

In mathematics, this is a row

❀❀❀❀❀❀❀❀❀❀❀❀❀❀❀

and this is a column.

❀
❀
❀
❀
❀
❀
❀

3.

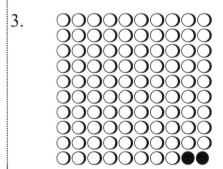

2%

4. To find $\frac{6}{7}$ of 413 you multiply 413 by 6 and then divide the result by 7.

$\frac{6}{7}$ of 413 is 354.

$$\begin{array}{r} 413 \\ \times\ \ 6 \\ \hline 2478 \end{array}$$

$$\begin{array}{r} 354 \\ 7\overline{)2478} \\ \underline{21}\ \ \ \\ 37\ \\ \underline{35}\ \\ 28 \\ \underline{28} \end{array}$$

Chapter Seven
A German

The paramedic went back to his fire truck. The policeman was busy directing traffic around the accident. Edward was attaching Minnie's car to his tow truck. The ambulance was leaving to take Minnie to the hospital.

Can you guess which five-year-old was forgotten in the midst of all the chaos? Hint: He teaches mathematics at KITTENS University and has a doll named Kingie.

Fred wandered over to the crowd that had gathered to look at the accident. Fred and Heidi recognized each other at the same time.

It was Heidi von Hilfe, the German teacher who has an office directly below Fred's. Many people know about her

because of the giant poster she has outside her office door.

Fred once thought about what it would be like to live in that German castle in the mist surrounded by a dark green forest. Certainly, it would be different than living in room 314 in the Mathematics Building on the KITTENS campus.

But there would be some drawbacks:
✓ Fred couldn't afford to buy it on a salary of $500/month.
✓ Even if he were given the castle as a gift, he couldn't afford the property taxes, the heating bill, the electric bill, the water

bill, and the upkeep (painting, plumbing, roof repair, vacuuming and dusting 50 rooms, etc.).

✓ It might get a little lonely with just Fred and Kingie in all those rooms.

✓ The commute each day from Germany to Kansas would be horrific. He would have to leave at midnight to get to his first class at 8 a.m. After his last class, which ends at 5 p.m., he would have to fly back to his castle. He would get home at 11 p.m. That would give him one hour to brush his teeth and sleep. Then at midnight he would do it all over again. The lack of sleep would make dark circles under his eyes. (Actually, the flight times would be longer than this.)

a very tired Fred if he commuted from Germany to Kansas every day

"Hello, Fred," said Heidi. "It looks like it was a pretty bad accident. Was it anyone we know?"

The KITTENS community is not that large. Some people know almost everyone in the area.

Fred told Heidi, "Minnie, from the blood bank, was the driver. She got a pretty severe bump on her head. They are taking her to the hospital to check it out."

"Oh, the poor dear. I hope she is all right.* We are often bridge partners at the bridge club on Wednesday nights. She sometimes makes some silly plays when she gets distracted trying to do too many different things at the same time."

When Heidi had asked, "Was it anyone we know?" Fred didn't get a chance to mention that he was also involved in the

---

* All right is sometimes misspelled as *alright.* A lot is sometimes misspelled as *alot.* Dumb is sometimes misspelled as *dum.*

All ready and already are both correctly spelled. They mean different things. "Are we all ready?" "He has already done that."

accident. Fred always told the truth. He knew that *leaving out* parts of the truth is not telling the whole truth.

"Hey. Do you play bridge?" Heidi asked. "Minnie isn't going to be at the bridge club tonight. I could use a partner."

Fred shook his head.

"Oh, you are so bashful. I'm sure you've played but are afraid to admit it."

Fred thought *I've seen kids dancing around in circles shouting "London bridges all fall down," but I didn't know that adults would meet every Wednesday night to play that.*

Fred didn't know that bridge was a card game. He had never even played *Fish* or *Old Maid*. He told Heidi, "I'm willing to learn. I've always liked singing and dancing."

Heidi thought that Fred's *I'm willing to learn* meant that he wasn't a very good player. What Fred meant was *I don't know how to play at all.*

Heidi thought that Fred's *I've always liked singing and dancing* was just a five-year-old's silly talk. She said, "You silly Ente," combining English and German. Fred knew what the German word *Ente* meant.

Heidi was on foot. She was not driving to the bridge club meeting since it was only six blocks from her house. She walked. When Fred saw the house they were heading to, he immediately knew who owned it.

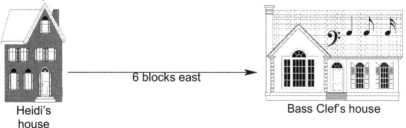

Heidi's house

6 blocks east

Bass Clef's house

B. Clef, the music professor, greeted them at the door.

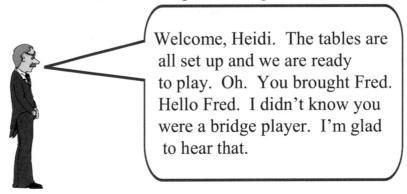

Welcome, Heidi. The tables are all set up and we are ready to play. Oh. You brought Fred. Hello Fred. I didn't know you were a bridge player. I'm glad to hear that.

Some brief notes . . .

♪#1: Professor Clef didn't notice Fred at first. All the other bridge players were at least five feet tall.

♪#2: Clef said, "I'm glad to *hear* that." He teaches music. Someone who teaches painting might have said, "I'm glad to *see* that." A mathematician might have said, "I *understand*." A wrestler, "I can *get a hold* of that."

♪#3: You might create a function whose domain is the set of all occupations and whose codomain is their favorite verb.

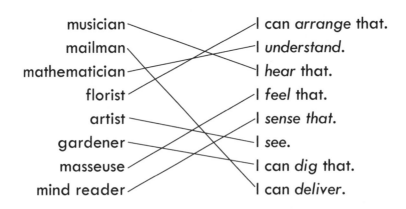

Domain	Codomain
musician	I can *arrange* that.
mailman	I *understand*.
mathematician	I *hear* that.
florist	I *feel* that.
artist	I *sense that*.
gardener	I *see*.
masseuse	I can *dig* that.
mind reader	I can *deliver*.

Domain   **Codomain**

Your Turn to Play

1. That German castle was built in the 1800s. If Fred owned the castle when it was new, could he have afforded the electricity bill at that time?

2. Suppose the lowest point in the castle was 1,859 feet above sea level, and the highest point was 2,307 feet above sea level. If Fred ran from the lowest point to the highest point, how many feet would he have climbed?

3. If he could climb at the rate of 4 feet/second, how long would it take him?

4. If Fred were living in that castle today and commuting to KITTENS University, he would get home every night at 11, and then at midnight he would have to get up to go back to KITTENS.

     In that one hour, he would spend 5 minutes brushing his teeth, 2 minutes changing into his nightclothes, 4 minutes saying his evening prayers, and after his evening's sleep, he would spend another 2 minutes changing into his teaching clothes.

     How long would he have to sleep?

5. Create a function whose domain is {Fred, Heidi, Bass} and whose codomain is {Texas, Iowa}.

Small Music Lesson

Fred knew that it was Bass Clef's house, because he had a bass clef 𝄢 printed on the roof of his house.

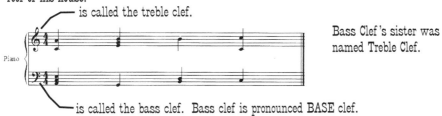

— is called the treble clef.

Bass Clef's sister was named Treble Clef.

— is called the bass clef. Bass clef is pronounced BASE clef.

. . . . COMPLETE SOLUTIONS . . . .

1. That many years ago, electric companies didn't exist.

    In 1925, there were 6,300,000 (six million, three hundred thousand) farms in the United States. Only 205,000 of them were receiving electricity from an electric company.

$$\begin{array}{r} 6300000 \\ -\ 205000 \\ \hline 6095000 \end{array}$$ 6,095,000 weren't.

2. $$\begin{array}{r} 2307 \\ -1859 \\ \hline 448 \end{array}$$ He would have to climb 448 feet.

3. The General Rule is that if you don't know whether to add, subtract, multiply, or divide, restate the problem with easier numbers and notice which operation you used.

    For example, how long would it take to climb 12 feet at the rate of 4 feet/second? It would take 3 seconds. You divided.

$$4\overline{)448}^{\,112}$$ It would take 112 seconds.

4. $$\begin{array}{r} 5 \\ 2 \\ 4 \\ +\ 2 \\ \hline 13 \end{array}$$ Of the 60 minutes (= 1 hour), he would spend 13 minutes not sleeping. That would leave 47 minutes for him to snooze. $$\begin{array}{r} 60 \\ -\ 13 \\ \hline 47 \end{array}$$

5. There are eight possible different functions that you could have created. Here's one of them: $F_{red} \rightarrow T_{exas}$, $H_{eidi} \rightarrow T_{exas}$, $B_{ass} \rightarrow I_{owa}$.

    Here's another: F→I, H→I, B→I.

        Or: F→T, H→I, B→I.

            Or: F→I, H→I, B→T.

Chapter Eight
Learning to Play Bridge

Everyone sat down. There were four people at each table. Heidi sat across from Fred. This confused him. If they were supposed to be partners, he wanted to sit next to her.

Fred also couldn't understand why everyone was sitting down. That seemed silly. If they were going to play London Bridges, they were supposed to be standing in a circle so that they could dance and sing and "all fall down."

Time Out!

Fred was totally confused. Since he had never had the chance to play with other children, he didn't even have the children's songs straight.

He thought it was called London Bridges. It was really London Bridge:

London Bridge is falling down,
Falling down, falling down.
London Bridge is falling down,
My fair lady.

London Bridge is Falling Down

Traditional

It is written in the treble clef, since that is the range in which most women and children sing. Men's voices are usually written in the lower range (bass clef).

Fred was also mixing up two different children's songs. The "all fall down" doesn't come from the London Bridge song.

It comes from

> Ring-a-round a rosie,
> A pocket full of posies,
> Ashes! Ashes!
> We all fall down.

At each table was a bowl of peanuts, a bowl of chocolate mints, and four glasses of Sluice. Everyone except Fred was busy eating and drinking. He imagined that all these bridge players must have skipped dinner. He didn't know that many people will eat whenever food is placed in front of them—whether they are hungry or not.

Bass Clef got three phone books for Fred to sit on so that he would be tall enough. Fred thanked him.

Heidi shuffled the cards. Maybe bridge is a card game Fred thought to himself. He gulped. He had never played any card games before. He hoped that the rules for bridge would be simple.

Heidi dealt out all 52 cards. Fred counted as she dealt them. He did the math:

$$\begin{array}{r} 13 \\ 4{\overline{\smash{\big)}\,52}} \\ \underline{4} \\ 12 \\ \underline{12} \end{array}$$

Each person got 13 cards.

The others picked up their cards and arranged them. They all had adult-sized hands. Holding 13 cards was not difficult.

Fred couldn't hold 13 cards in one hand. His hands were too small. If I hold 5 cards in my left hand, then I will have to hold 8 cards in my right hand. (13 - 5 = 8)

If I hold 7 cards in my left hand, then I will have to hold 6 cards in my right hand. (13 - 7 = 6)

He realized that no matter how he arranged them, one of his hands would have to hold at least seven cards.

← Fred's hand

This is silly. I can only hold one card in each hand.

Fred did the only logical thing. He picked up each card, memorized it, and then put it down.

It turned out that memorizing the 13 cards he had been dealt wasn't very difficult.

Fred did not understand what A, J, Q, and K meant. He also did not realize that this was a "dream hand" (a hand that that someone playing bridge all their life might never see).

"Aren't you going to pick up your cards?" Heidi asked.

"I'm okay," Fred said.

The player on Heidi's left said, "One no trump."

Everyone looked at Fred.

Small drops of sweat formed on Fred's forehead. Apparently, he was supposed to say something. The only clue was that a player said, "One no trump." Fred did not have any idea what that meant.

Suddenly, it seemed clear. What that player meant was "One no Donald Trump."

So Fred said, "Two no Walt Disney."

Everyone was silent. Everyone was looking at Fred. Heidi began to giggle. She whispered to him, "Look at your cards and say 'two' of whatever suit you have the most of."

Fred said, "Two puppy paws."

That was a very natural mistake.

The next player said, "Two diamonds."

Heidi said, "I'll pass."

The player on Heidi's left said, "Pass."

Fred said, "Two and one-half puppy paws."

Heidi explained to him that puppy paws are called clubs and that his bid had to be a natural number.

Fred knew what the set of natural numbers is: {1, 2, 3, 4, 5, 6, 7, 8, 9, 10, 11, 12 . . .}.

So he said, "Three clubs."

Everyone then said, "Pass."

Heidi told Fred that he had "won the bid."

This is the easiest game in the world Fred thought. All you have to do is name the biggest number. It is surprising that no one at the table could think of a natural number larger than three.

Fred started to gather together his cards so that they could play another time. He thought that the game was over.

Heidi told him to play a card. He picked a card at random and put it in the center of the table. It was a 🃞.

Everyone else tossed in a card, and Heidi shoved the four cards to Fred and said, "You won the trick. Play another card."

Fred didn't know what a *trick* was, but he didn't care.

He played another card and won another trick. He won all 13 tricks.

Fred could not imagine why adults spend an evening playing bridge. It seemed like a stupid game. It was so easy to win.*

Your Turn to Play

1. If you dealt out a 52-card deck to 3 people, would it come out even or would there be one or more cards left over?
2. The even natural numbers are 2, 4, 6, 8, 10, 12, 14. . . .
The odd natural numbers are 1, 3, 5, 7, 9, 11, 13, 15. . . .

    If you were dealt 13 cards and you wanted to hold some of the cards in each hand . . .

    A) Could you have an *even* number of cards in each hand?

    B) Could you have an *odd* number of cards in each hand?
3. Is an odd natural number times an odd natural number always odd?
4. Is an even natural number times an even natural number always even?
5. What is $\frac{5}{13}$ of 52?

---

* The truth is that bridge is quite a difficult game to learn to play well. It takes years of practice. Fred won easily because he had been dealt 13 puppy paws—I mean 13 clubs.

.......COMPLETE SOLUTIONS.......

1. When Fred wanted to find out how 52 cards would be divided up into 4 equal groups, he wrote

$$4\overline{)\,52}\;\;^{13}$$

We want to divide 52 cards among 3 people.*

So we divide 3 into 52.

$$3\overline{)\,52}\;\;^{17\ R\ 1}$$
$$\underline{3}$$
$$22$$
$$\underline{21}$$
$$1$$

It would not come out even. Each person would receive 17 cards with one left over.

2A. No. If you have an *even* number of cards in each hand, then the sum will always be even. **For example, 2 + 6 = 8.**

2B. No. If you have an *odd* number of cards in each hand, then the sum will always be even. **For example, 7 + 13 = 20.**

3. You have to try out a bunch of numbers to see if an odd number times an odd number is always odd. After you've done a couple minutes' worth, you start to believe it is. And you are right.

4. An even number times an even number is always even.

5. $\frac{5}{13}$ of 52 means multiply 52 times 5 and then divide by 13.

$$52$$
$$\underline{\times\ 5}$$
$$260$$

$$13\overline{)\,260}\;\;^{20}$$
$$\underline{26}$$
$$00$$

$\frac{5}{13}$ of 52 is 20.

---

* English lesson time! You say *among* three people or *among* four people, but you say *between* two people.

Chapter Nine
Doing or Watching

Three of the four glasses of Sluice at Fred's table were empty. Originally there had been 7,648 peanuts in the bowl of peanuts. Seven-eighths ($\frac{7}{8}$) of them had been eaten.

$\frac{7}{8}$ of 7,648 had been eaten.

7648 times 7 and divide by 8.

$$
\begin{array}{r}
7648 \\
\times\ 7 \\
\hline
53536
\end{array}
$$

$$
\begin{array}{r}
6692 \\
8\,\overline{)\,53536} \\
\underline{48} \\
55 \\
\underline{48} \\
73 \\
\underline{72} \\
16 \\
\underline{16}
\end{array}
$$

6,692 peanuts had been eaten.

small essay

Two Kinds of People

There are those that *do* things, and there are those that watch other people do things.

Those people that watch others are called *critics*. Being a critic is much easier than doing things yourself.

Today, you get to be the critic. On this page I did the work of finding out what $\frac{7}{8}$ of 7,648 is. Would you please carefully look over my work and check to make sure that I did it right?

end of small essay

Fred, of course, didn't eat any of the peanuts. He had had a big dinner with Minnie when they ate at *K's Afternoon Dining*. (Fred had a sip of water, and that was plenty for him.)

Small timeout: *big dinner* is an example of irony. Irony is saying the opposite of what you really mean—when both you and the reader know that you are kidding. Irony is one way of emphasizing something.

Children's books have very little irony compared with adult literature. Kids might not realize that you are obviously kidding when you say that Fred had a big dinner.

Could the three other card players divide the 6,692 peanuts equally? (Be a critic again and check my work.)

$$
\begin{array}{r}
2230\ \text{R}2 \\
3\overline{)6692} \\
\underline{6} \\
06 \\
\underline{6} \\
09 \\
\underline{9} \\
02 \\
\underline{0} \\
2
\end{array}
$$

This is called long division.

They each had 2,230 peanuts with 2 left over. So they could not divide the peanuts equally.

Wait a minute! I, your reader, have a question. I'm going to interrupt your story.

If that is called long division, is there such a thing as short division?

Yes there is, but very few math books ever talk about it.

Could you talk about it while all those bridge players are busy munching on their stupid peanuts?

I would be glad to. Thank you for asking. Short division is a lot faster than long division, but there is only one drawback.

I knew there would be a drawback. What is it?

You can only use it when you have a one-digit divisor.

5 has one digit.

392 has three digits.

773,980 has six digits.

$$\frac{\text{quotient}}{\text{divisor})\ \text{dividend}}$$

Let's do a long division problem by short division.

$$\begin{array}{r} 23 \\ 4)\overline{92} \\ \underline{8} \\ 12 \\ \underline{12} \end{array}$$ becomes $4)\overline{9^12}$ with 23 on top   It's shorter!

Here's how it works:

4 goes into 9 twice. Put a 2 on top.

2 times 4 is 8. Subtract 8 from 9.

Put a tiny 1 up near the 2.

4 goes into 12 three times. Put a 3 on top.

Here is the long division from two pages ago:

$$\begin{array}{r} 6692 \\ 8)\overline{53536} \\ \underline{48} \\ 55 \\ \underline{48} \\ 73 \\ \underline{72} \\ 16 \\ \underline{16} \end{array}$$ becomes $8)\overline{53^53^73^16}$ with 6692 on top   Much shorter!

Hey! I like short division.

Bass Clef announced that his sister Treble had made a
chocolate cake for the bridge club.

He also made some ice cream sundaes to go with it.

Everyone took a break, stood up, and headed toward the
kitchen. They had been playing cards for one-quarter of an
hour.

(Continue to be a critic and watch me do the math.)

One-quarter of an hour means . . .

$\frac{1}{4}$ of an hour means . . .

$\frac{1}{4}$ of 60 minutes means . . .

60 minutes multiplied by 1 and divided by 4 which is . . .

$$
\begin{array}{c}
60 \\
\times\ 1 \\
\hline
60
\end{array}
\qquad
\begin{array}{r}
15 \\
4\overline{)60} \\
4 \\
\hline
20 \\
\underline{20}
\end{array}
\quad \text{or} \quad
\begin{array}{r}
15 \\
4\overline{)60}
\end{array}
$$

So one-quarter of an hour is 15 minutes.

One of the bridge players took Fred's full
glass of Sluice, drank a little of it, and put some
ice cream into the glass to make an ice cream
float. (It's also called an ice cream soda.)

Fred couldn't understand it. After 15
minutes of card playing, everyone was ingesting
4,000 Calories.

Everyone except Fred. He was watching.
A skinny critic.

Your Turn to Play

1. Divide, using short division $7\overline{)574}$

2. Divide, using short division $9\overline{)578}$

3. A normal person might eat about 2,000 Calories each day. How many day's worth of food was this 4,000-Calorie snack at the bridge club?

4. How many minutes in two-fifths ($\frac{2}{5}$) of an hour?

5. Bass took an even number of peanuts out of a bowl and put some of them in each hand. If he had an odd number of peanuts in his right hand, what can you say about the number of peanuts in his left hand?

6. A **secant line** to a circle is a (straight) line that hits the circle in two places.

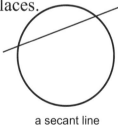

a secant line

If Bass cut Treble's cake with 3 secant lines, what is the most pieces he could make? (The answer is more than six.)

7. Using 3 secant lines, can he make 5 pieces?

8. Using 3 secant lines, can he make 4 pieces?

·······**COMPLETE SOLUTIONS**·······

1.
$$7\overline{)57^14}$$
with 82 on top

2.
$$9\overline{)57^38}$$
with 64 R2 on top

3. If 2,000 Calories is one day's worth, then a 4,000-Calorie snack is two days' worth of food.

4. $\frac{2}{5}$ of an hour

  $\frac{2}{5}$ of 60 minutes

  60 multiplied by 2 and divided by 5

$$\begin{array}{r} 60 \\ \times\ 2 \\ \hline 120 \end{array}$$ $$5\overline{)1\,2^20}$$ with 24 on top   Two-fifths of an hour is 24 minutes.

5. He must have had an odd number in his left hand. An odd number plus an odd number always equals an even number.

6. Here's a way to make 7 pieces with 3 cuts:

7. Five pieces with 3 cuts: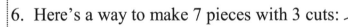

8. Four pieces with 3 cuts: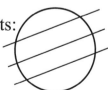

Chapter Ten
In the Cool Evening Air

The bridge club meeting started at 7 p.m. At 7:15 they took a snack break. (We are not counting the bowl of peanuts, the bowl of chocolate mints, and the glasses of Sluice at each table.)

By 7:30 the cake and the ice cream sundaes were all consumed. People were putting on their coats and leaving. Fred thanked Bass for his hospitality.

Bass told Fred, "You are sure a fine bridge player. I do hope that you come back next Wednesday."

Fred didn't know what to say. He smiled and remained silent.

Outside, he thanked Heidi for inviting him to the bridge club* and started walking back to the Math Building on campus. The air was starting to cool off a bit.

Fred realized that he liked it both ways: time with other people and time alone. On the other hand, it would be terrible to always be in a crowd . . . and it would be terrible to always be alone.

---

* Would you like people to like you? Learn to thank people. Even God likes to be thanked.

As Fred walked across the Great Lawn on the KITTENS campus, he thought about Minnie. He hoped that she was okay. He remembered telling the paramedic that about two percent of those admitted to a hospital as a result of a blow to the head experience an extradural hemorrhage.

He suddenly realized the obvious: (1) she had had a blow to her head and (2) she had been admitted to the hospital.

Two percent (2%) lit up in Fred's brain. It wasn't just a number anymore. It had a meaning.

He turned and started walking toward KITTENS hospital. He needed to see how Minnie was doing.

2% means 2 parts out of a hundred.

2% is the same as the fraction $\frac{2}{100}$

3% is the same as the fraction $\frac{3}{100}$

9% is the same as the fraction $\frac{9}{100}$

50% is the same as the fraction $\frac{50}{100}$

The percent sign (%) means "parts per hundred."

84% is the same as the fraction $\frac{84}{100}$

96% is the same as the fraction $\frac{96}{100}$

Multiple choice question. Pick the correct answer.

Changing percents into fractions is

    A) hard

    B) really easy

If you picked A) hard then do this homework assignment.   (If you selected B) really easy, then you can skip this.)

This stuff is called DRILL-AND-KILL.

$5\% = \frac{?}{100}$	$62\% = \frac{?}{100}$	$80\% = \frac{?}{100}$	$56\% = \frac{?}{100}$
$7\% = \frac{?}{100}$	$76\% = \frac{?}{100}$	$11\% = \frac{?}{100}$	$72\% = \frac{?}{100}$
$3\% = \frac{?}{100}$	$37\% = \frac{?}{100}$	$46\% = \frac{?}{100}$	$24\% = \frac{?}{100}$
$55\% = \frac{?}{100}$	$93\% = \frac{?}{100}$	$23\% = \frac{?}{100}$	$42\% = \frac{?}{100}$
$86\% = \frac{?}{100}$	$38\% = \frac{?}{100}$	$62\% = \frac{?}{100}$	$22\% = \frac{?}{100}$
$87\% = \frac{?}{100}$	$81\% = \frac{?}{100}$	$89\% = \frac{?}{100}$	$75\% = \frac{?}{100}$
$23\% = \frac{?}{100}$	$40\% = \frac{?}{100}$	$77\% = \frac{?}{100}$	$55\% = \frac{?}{100}$
$6\% = \frac{?}{100}$	$50\% = \frac{?}{100}$	$59\% = \frac{?}{100}$	$29\% = \frac{?}{100}$
$77\% = \frac{?}{100}$	$69\% = \frac{?}{100}$	$87\% = \frac{?}{100}$	$44\% = \frac{?}{100}$
$14\% = \frac{?}{100}$	$60\% = \frac{?}{100}$	$65\% = \frac{?}{100}$	$32\% = \frac{?}{100}$
$32\% = \frac{?}{100}$	$17\% = \frac{?}{100}$	$41\% = \frac{?}{100}$	$48\% = \frac{?}{100}$
$43\% = \frac{?}{100}$	$1\% = \frac{?}{100}$	$57\% = \frac{?}{100}$	$53\% = \frac{?}{100}$
$99\% = \frac{?}{100}$	$37\% = \frac{?}{100}$	$8\% = \frac{?}{100}$	$77\% = \frac{?}{100}$
$48\% = \frac{?}{100}$	$58\% = \frac{?}{100}$	$74\% = \frac{?}{100}$	$49\% = \frac{?}{100}$
$65\% = \frac{?}{100}$	$33\% = \frac{?}{100}$	$68\% = \frac{?}{100}$	$34\% = \frac{?}{100}$
$45\% = \frac{?}{100}$	$98\% = \frac{?}{100}$	$43\% = \frac{?}{100}$	$51\% = \frac{?}{100}$
$51\% = \frac{?}{100}$	$21\% = \frac{?}{100}$	$83\% = \frac{?}{100}$	$39\% = \frac{?}{100}$

Here are the answers so that you can check your work.

$5\% = \dfrac{5}{100}$ $62\% = \dfrac{62}{100}$ $80\% = \dfrac{80}{100}$ $56\% = \dfrac{56}{100}$

$7\% = \dfrac{7}{100}$ $76\% = \dfrac{76}{100}$ $11\% = \dfrac{11}{100}$ $72\% = \dfrac{72}{100}$

$3\% = \dfrac{3}{100}$ $37\% = \dfrac{37}{100}$ $46\% = \dfrac{46}{100}$ $24\% = \dfrac{24}{100}$

$55\% = \dfrac{55}{100}$ $93\% = \dfrac{93}{100}$ $23\% = \dfrac{23}{100}$ $42\% = \dfrac{42}{100}$

$86\% = \dfrac{86}{100}$ $38\% = \dfrac{38}{100}$ $62\% = \dfrac{62}{100}$ $22\% = \dfrac{22}{100}$

$87\% = \dfrac{87}{100}$ $81\% = \dfrac{81}{100}$ $89\% = \dfrac{89}{100}$ $75\% = \dfrac{75}{100}$

$23\% = \dfrac{23}{100}$ $40\% = \dfrac{40}{100}$ $77\% = \dfrac{77}{100}$ $55\% = \dfrac{55}{100}$

$6\% = \dfrac{6}{100}$ $50\% = \dfrac{50}{100}$ $59\% = \dfrac{59}{100}$ $29\% = \dfrac{29}{100}$

$77\% = \dfrac{77}{100}$ $69\% = \dfrac{69}{100}$ $87\% = \dfrac{87}{100}$ $44\% = \dfrac{44}{100}$

$14\% = \dfrac{14}{100}$ $60\% = \dfrac{60}{100}$ $65\% = \dfrac{65}{100}$ $32\% = \dfrac{32}{100}$

$32\% = \dfrac{32}{100}$ $17\% = \dfrac{17}{100}$ $41\% = \dfrac{41}{100}$ $48\% = \dfrac{48}{100}$

$43\% = \dfrac{43}{100}$ $1\% = \dfrac{1}{100}$ $57\% = \dfrac{57}{100}$ $53\% = \dfrac{53}{100}$

$99\% = \dfrac{99}{100}$ $37\% = \dfrac{37}{100}$ $8\% = \dfrac{8}{100}$ $77\% = \dfrac{77}{100}$

$48\% = \dfrac{48}{100}$ $58\% = \dfrac{58}{100}$ $74\% = \dfrac{74}{100}$ $49\% = \dfrac{49}{100}$

$65\% = \dfrac{65}{100}$ $33\% = \dfrac{33}{100}$ $68\% = \dfrac{68}{100}$ $34\% = \dfrac{34}{100}$

$45\% = \dfrac{45}{100}$ $98\% = \dfrac{98}{100}$ $43\% = \dfrac{43}{100}$ $51\% = \dfrac{51}{100}$

$51\% = \dfrac{51}{100}$ $21\% = \dfrac{21}{100}$ $83\% = \dfrac{83}{100}$ $39\% = \dfrac{39}{100}$

Hee! Hee! Hee! Hee! Hee! Hee! Hee! Hee! Hee! Hee! **I have just been playing these last two pages.**

Unless you thought that $21\% = \frac{21}{100}$ was hard, you didn't need to do a million problems. In contrast to the previous two pages of mindless repetitive drill, the problems in the *Your Turn to Play* are much fewer and less boring.

Your Turn to Play

(One question!)

1. Approximately 2% of those admitted to a hospital as a result of a blow to the head will experience an extradural hemorrhage.

This kind of brain bruise is more serious than a bruised knee, *but it doesn't mean that the patient is going to die.*

Suppose 550 people are admitted to the hospital because of trauma to the head. (*Trauma*—rhymes with *drama*—is the fancy medical word for injury.)

Here are 550 people with head injuries so that you can get an idea of how many 550 are:

☺☺
☺☺
☺☺
☺☺
☺☺
☺☺
☺☺
☺☺
☺☺
☺☺
☺☺☺☺☺☺☺☺☺☺☺☺☺☺☺☺☺☺☺☺☺☺☺☺☺☺☺☺☺☺☺☺☺☺☺☺☺.

How many of these people would you expect to have extradural hemorrhages?

In other words, what is 2% of 550?

·······COMPLETE SOLUTION·······

1.   2% of 550

$\frac{2}{100}$ of 550

550 multiplied by 2 and divided by 100

$$
\begin{array}{r}
550 \\
\times \ 2 \\
\hline
1100
\end{array}
\qquad
\begin{array}{r}
11 \\
100\overline{)1100} \\
\underline{100} \ \ \\
100 \ \ \\
\underline{100} \ \ \\
\end{array}
$$

Of those 550 people,

☺☺
☺☺
☺☺
☺☺
☺☺
☺☺
☺☺
☺☺
☺☺
☺☺☺☺☺☺☺☺☺☺☺☺☺☺☺☺☺☺☺☺☺☺☺☺☺☺

we might expect that 11 of them would have extradural hemorrhages. ⊗⊗⊗⊗⊗⊗⊗⊗⊗⊗⊗.

That's not many people. Extradural hemorrhages are rare.

Chapter Eleven
Minnie

Fred started jogging toward the hospital. He had been sitting too long and needed the exercise. Young boys feel better and are healthier if they get a reasonable amount of exercise.*

He had read in one of his medical books that there are three things to avoid: Sitting, Sipping, and Smoking.

Sitting means not getting enough exercise.

Sipping means drinking too much alcohol.

Smoking means . . . smoking.

Fred walked in the main entrance to the hospital. He asked at the information desk about Minnie.

The receptionist said that Minnie was doing fine. She was in room 305 on the third floor.

Fred thanked her and headed up the stairs.

---

* That is also true for young girls, middle-aged men, middle-aged women, old men, old women, and pet dogs. In contrast, putting your bowling ball in front of the television all day long will not affect its health at all.

On the door to the stairs was an official hospital sign:

Do not use
these stairs
if you are
in a wheelchair!

*otherwise always use them
instead of the elevator!
It's better for you.*

Fred thought that was a silly sign. He noticed that someone had scribbled a graffito* at the bottom of the sign. Fred stood close to the sign and read: *otherwise always use them instead of the elevator! It's better for you.*

He ran up the two flights of stairs and walked down the hallway to room 305. Minnie was sitting up in bed. She looked a little different than when Fred first met her earlier this afternoon. She had a bandage on her head, and she wasn't wearing her earrings. The television was on and she was reading a newspaper.

"Hi, Fred," she said. "You little bunny. How are you? Did you get hurt in the accident?"

---

* Graffiti are signs, markings, drawings that people illegally make on buildings, bathroom walls, sidewalks, etc. Rarely do you see just one example of graffiti appearing alone. That's why *graffito* is such a rare word.

One cow, two cows, three cows.

One graffito, two graffiti, three graffiti.

Minnie was the only one in the world who called Fred, "little lamb," "smart little monkey," and "little bunny." Everyone else called him either Fred or Professor Gauss. Minnie liked metaphors.

Fred said, "I'm fine. I was wearing seatbelts, and my head didn't get bonked. Have the doctors told you anything about your condition?"

"I feel okay except for a little headache. The docs want me to stay here overnight just to make sure I don't have a *brain bruise*. That's the silly name he called it. I phoned my husband, Karl, and he will be here any moment. The only thing I don't know about is my car."

Fred shrugged his shoulders. "I'm afraid that you and Karl will have to go shopping for another car."

Karl arrived and rushed to her bedside. Fred quietly excused himself and left the room. The last thing he heard Minnie say was, "Hello, you big teddy bear."

picture of
Karl
at the
beach

As Fred walked down the hallway he wondered if someday he would graduate from "*little* bunny" to "*big* teddy bear."

He recalled that Minnie wasn't the only one to give him a pet name*. Heidi had called him a "silly Ente."

---

* A pet name means a name you give a person whom you like. If you had a dog named Ralph and you called him Ralphie, that would be a pet pet name. Ralph is your pet's name.

It wasn't even eight o'clock yet. Fred decided to do something a little different. The KITTENS University Gym was open until 10 every night. Maybe it's time I do more than just jogging Fred thought. Perhaps a little working out in the gym might be what I need.

In truth, Fred wanted to look a little more like Karl.

The KITTENS University Gym is the newest building on campus.*

Fred headed inside. There were many different areas of the gym:

☞ Rows of treadmills. Fred didn't need that. His heart and lungs were in great condition from all the years of jogging.

☞ Dance and stretching classes. Like most five-year-olds, Fred was very flexible. He could easily put his leg behind his head.

☞ The snack bar. One of Fred's students named Joe was there. He was having a jelly-bean sandwich.

☞ The weight room. Yes!

Just looking at the door told Fred that this was the right place for him.

---

* Several months ago the only local gym (Coalback Gym, 789 Main Street) suddenly went out of business (in *Life of Fred: Dogs*).

Fred pushed on the door to the weight room. It wouldn't open.

---

1. Joe's jelly-bean sandwich weighed 20 ounces. Three-fifths ($\frac{3}{5}$) of the weight of the sandwich was sugar.

How many ounces of sugar were in the sandwich?

(In other words, what is $\frac{3}{5}$ of 20?)

2. If Minnie had been in room 578, what floor would that have been on?

3. In the last hour (60 minutes) Fred has spent 40% of that time watching people eat. How many minutes is that?

(In other words, what is 40% of 60 minutes?)

4. $74\% = \frac{x}{100}$ What does x equal?

5. Joe was eating jelly beans at the rate of 8 per second. How many would he eat in 88 seconds?

.......COMPLETE SOLUTIONS.......

1. Three-fifths of a jelly-bean sandwich is sugar

   $\frac{3}{5}$ of 20 ounces

   20 multiplied by 3 and divided by 5

   $$\begin{array}{r} 20 \\ \times\ 3 \\ \hline 60 \end{array} \qquad \begin{array}{r} 12 \\ 5\overline{)6\,0} \end{array} \text{ (using short division)}$$

   There are 12 ounces of sugar in a jelly-bean sandwich.

2. Room <u>5</u>78 is usually on the fifth floor.

3. Forty percent of one hour

   40% of 60 minutes

   $\frac{40}{100}$ of 60

   60 times 40 divided by 100

   $$\begin{array}{r} 60 \\ \times 40 \\ \hline 00 \\ 240\ \ \\ \hline 2400 \end{array} \qquad \begin{array}{r} 24 \\ 100\overline{)2400} \\ 200\ \ \\ \hline 400 \\ 400 \\ \hline \end{array}$$

   Forty percent of one hour is 24 minutes.

4. $74\% = \frac{74}{100}$ so x = 74

5. Joe is eating at the rate of 8 per second for 88 seconds. The General Rule is if you don't know whether to add, subtract, multiply, or divide, restate the problem using easier numbers and notice which operation you used. If Joe were eating at the rate of 3 jelly beans per second and he ate for 2 seconds, he would have eaten 6 jelly beans. We multiplied.

   $$\begin{array}{r} 88 \\ \times\ 8 \\ \hline 704 \end{array}$$ Joe would eat 704 jelly beans in 88 seconds.

Chapter Twelve
Weight Room

Fred pushed on the door to the weight room. It wouldn't budge. He checked to make sure there wasn't a door knob to turn. There wasn't. He pushed with all of his might.

Fred thought I must be really weak. I can't even open the door to the weight room. Is this the way that they keep really little guys like me from getting into the weight room?

Then someone came out of the weight room. She had no trouble opening the door.

Fred had been trying to *push* on a door when he should have *pulled*. He was embarrassed.

He would have been more embarrassed if he had been pushing on the door when she opened it. She probably would have knocked him down. Thirty-seven pounds (= one Fred) is not considered a large weight for a barbell in a weight room.

lbs. means pounds

Fred entered the weight room. He made a mental note that when he *left* the room he should push the door instead of

pulling it. He told himself When I enter, I should pull. When I exit, I should push. He didn't want any more embarrassment.

There were tons of barbells and dumbbells around the perimeter of the room. Fred had never seen them before. He guessed that barbells were for the big people and dumbbells were for kids. (He was wrong.)

One woman was using two dumbbells. He watched her to learn how to exercise with the dumbbells. When he first saw the weights, he thought that all you were supposed to do was pick them up and carry them around the room. (He was wrong.) She did a lot of different things with the dumbbells.

When she set the weights down, Fred was ready to go into action. He could feel his muscles growing even before he picked up his first weight.

The only problem was . . . picking up the weight. More embarrassment.

Maybe it's time I rest for a while Fred thought. He saw other weight lifters sitting on benches resting. The only difference is that my shirt isn't drenched in sweat like theirs. In truth, no one in the room was watching Fred. He could have been wearing a clown costume and no one would have noticed.

He went and sat in a corner and did some thinking. He decided to play one of his favorite games: Create-a-Function. This time he would think of functions where the domain and codomain were sets of things at the gym.

> ## A Small Review of Functions (part 1)
>
> You start out by naming two sets. Call the first set the domain and the second set the codomain.

For his first function, Fred decided that the first set (domain) would be the set of all people in the gym right now. The second set (codomain) would be the rooms in the gym.*

> ## A Small Review of Functions (part 2)
>
> Definition: A function is a rule which assigns to each member of the domain exactly one member of the codomain.

Fred created a function (a rule): *Assign each person to the room they are now in.*

Was this a function? Each person in the gym was assigned to exactly one room in the gym, namely the room that he or she was in right now. Yes, it is a function.

When Fred taught about functions in his math classes, he liked to say that if you can count up to the number one, you can tell whether a rule is a function. Check to see if each member of the first set (domain) has exactly one image in the second set.

The woman with the two dumbbells was assigned to the weight room. Joe was assigned to the snack bar room. Fred was assigned to the weight room. Each person was assigned to exactly one room.

_____

* {weight room, treadmill room, dance room, snack bar room, locker room}

Fred created a second function. Starting with the same domain (all the people in the gym) and the same codomain (all the rooms in the gym), Fred's new rule was: Assign everybody to the treadmill room.

Was this a function? Yes. Each element of the domain has exactly one image. The image of Fred is the treadmill

room. The image of is the treadmill room. The image of Joe is the treadmill room.

Some rules are *not* functions. How about the rule: *Assign each person to the room they were in an hour ago?*

Certainly, no person under that rule would be assigned to two different rooms, but under that rule some people would not be assigned to any member of the codomain. Take, for example, Fred. He wasn't in the gym an hour ago. He would not be assigned to any member of the codomain. That rule is not a function.

What about the rule: *Assign to each person to the room that is their favorite room?*

Joe's favorite room is the snack room.

has two favorite rooms: the weight room and the treadmill room. This is not a function. Each member of the domain must have exactly one image—not two images and not zero images.

When we look to see if a rule is a function . . .
☞ we don't care if some member of the codomain is hit twice (is the image of two different members of the domain).

☛ we don't care if some member of the codomain isn't hit at all (isn't the image of any member of the domain).

All we have to look at are the members of the first set (domain) and make sure each of them has exactly one image in the second set (codomain).

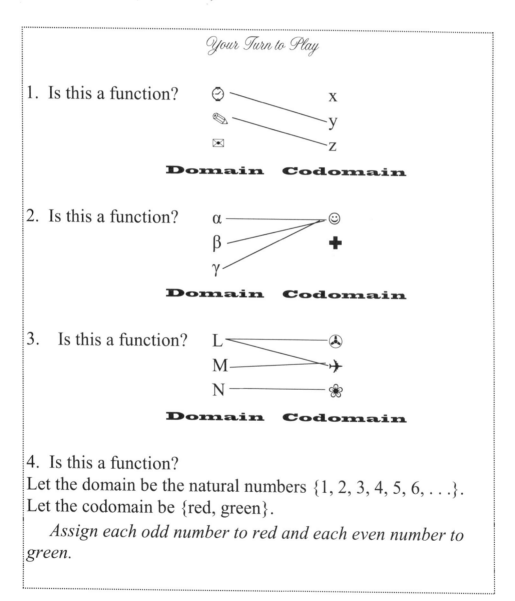

Your Turn to Play

1. Is this a function?

    Domain   **Codomain**

2. Is this a function?

    Domain   **Codomain**

3. Is this a function?

    Domain   **Codomain**

4. Is this a function?
Let the domain be the natural numbers {1, 2, 3, 4, 5, 6, . . .}.
Let the codomain be {red, green}.

   Assign each odd number to red and each even number to green.

```
. . . . . . . COMPLETE SOLUTIONS . . . . . . .
```

1. ☺ was assigned one image.

    ✎ was assigned one image.

    ✉ was not assigned an image. It's not a function.

2. α   was assigned one image.

    β   was assigned one image.

    γ   was assigned one image. Yes, it is a function.

3. L was assigned two images. It is not a function.

4. Will every member of {1, 2, 3, 4, . . . } be assigned *at least* one image? Yes, since every natural number is either odd or even.

    Will every member of {1, 2, 3, 4, . . . } be assigned *at most* one image? Yes, since no natural number can be both odd and even at the same time.

    This is a function.

It's been a long time since we've done a Row of Practice. Cover the gray answers with a blank sheet of paper.

<u>A Row of Practice</u>. *Do the whole row before you look at the answers.*

75	507	46	94	867
+ 87	− 9	× 36	× 7	+ 458
162	498	276	658	1325
		138		
		1656		

Chapter Thirteen
Pulling and Pushing

Fred's mind was clearer now. It helped to take a break. Sitting in a corner and thinking about functions took away a lot of the stress of being in the weight room. He realized that picking up those heavy weights wasn't the place for him to start his workout.

Instead, of *lifting*, he decided to *pull down*. After the guy with the big muscles finished with the pull-down machine, Fred knew exactly what he was going to do.

First, he would get rid of the heavy weights. The big guy helped him remove the weights that Fred couldn't lift.

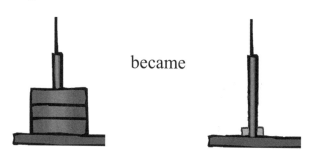

became

Fred was now ready for a serious workout. He carefully positioned himself. He knew that the correct form is very important.

He sat up straight.

He took a firm, but not too tight, grip on the rope.

He exhaled as he smoothly pulled on the rope.

Fred decided that he would build up his muscles some other day.

The big guy put Fred back on the ground.

Fred remembered to push on the door as he left the weight room. He didn't need to head to the locker room to shower up after his workout. None of the other places in the gym—treadmills, dance classes, snack bars—appealed to him.

He headed outside. It was 8:30. The sky wasn't dark yet because it was summertime. He didn't need to head back to his office to prepare for classes. The university president had shut down classes for the next eight days.

He wasn't hungry. (Some people eat because they are bored, not because they are hungry.)

He didn't need to shop for anything. (Some people go shopping because they are bored, not because they need something.)

He could have stayed in the gym and watched television for an hour or two, but he didn't want to. (Some people watch mindless television just to pass the hours.)

Fred walked down one of the many paths through the campus. He came to the edge of the rose garden and sat on one of the park benches. His feet did not touch the ground.

Most of the students and faculty had left KITTENS when the classes were cancelled. It was quiet. Fred thought of the words that Carlyle had written in *Sartor Resartus*:

Silence is the element
in which
great things
fashion themselves.

Some of Fred's students never seem to have a moment of silence. From the moment they wake up until they go to sleep at night, their ears never get a rest. One of his students had bragged to Fred that his little device that he had plugged into his ear held 3,982,072,442,131,552,101,307 songs. He never turned it off, even when he was attending one of Fred's classes.

Fred tried to picture what it would be like to have a drummer permanently parked in your head.

There is a time to be with people and a time to be alone. Fred liked both.

There is a time to direct your thoughts (such as when you are working on a math problem: $62\% = \dfrac{62}{100}$) and there is a time to not direct your thoughts.

Not directing your thoughts is a little like dreaming while you are awake. You watch "the show" rather than run the show. You let go.

Fred just sat there. He felt a warm breeze. He smelled the roses from the garden. The events of the last four hours poured over him . . . meeting Miss Ente at the bus stop, unloading the dozen boxes, looking forward to nine days at Camp Horsey-Ducky, the real fenders on the camp car, being towed down the road with a rope around the radiator, the lion doll at the blood bank, playing bridge. . . .

Random thoughts popped up . . . a giant ice cream machine, honey cards to practice multiplication, the piano in Kingie's fort, all the books that Prof. Eldwood has written, how Bass Clef had captured Coalback, a pain in my tummy,* the German castle surrounded by a dark green forest. . . .

All these are called brain chatter. When you first become quiet, brain chatter is the first thing that hits you.

It is the stillness that comes (minutes? hours?) after the brain chatter dies down that is the place in which great things fashion themselves.

Fred did not get to that stillness place this evening. As he looked out over the rose garden, he couldn't help noticing that a deer was eating the rose bushes.

---

* Fred would often get a pain in his tummy. He had never learned that when you don't eat for a day or two, your stomach says, "It would be nice if you sent some food down here!"

No one had ever told Fred that that feeling is called hunger—or what can be done to stop being hungry. Fred never figured out why people often eat several times each day.

1. There are 8 patches of roses in the KITTENS rose garden. The deer had eaten 2 of those 8 patches.

In the language of fractions, he had eaten $\frac{2}{8}$ of the roses.

What fraction of the roses had he *not* eaten?

2. Two-eighths ($\frac{2}{8}$) of a piece of pie is the same as one-fourth ($\frac{1}{4}$) of a piece of pie. Take a look at these two pictures:

Changing $\frac{2}{8}$ into $\frac{1}{4}$ is called **reducing the fraction**.

To reduce a fraction, you divide the top and bottom of the fraction *by the same number*.

Example: $\frac{9}{12} = \frac{3}{4}$ I divided the top and bottom by 3.

Your turn: reduce the fraction $\frac{10}{15}$

┌───┐

. . . . **COMPLETE SOLUTIONS** . . . .

1.  $\dfrac{6}{8}$ of the rose patches had not been eaten.

2. Dividing the top and bottom by 5,

$$\dfrac{10}{15} \text{ will equal } \dfrac{2}{3}$$

└───┘

$\dfrac{2}{3}$ is a lot easier to work with than $\dfrac{10}{15}$

Suppose someone asked you to find $\dfrac{10}{15}$ of 60. You know how to do that. 60 times 10 and divide the answer by 15.

$$
\begin{array}{cc}
\begin{array}{r} 60 \\ \times\,10 \\ \hline 600 \end{array}
&
\begin{array}{r} 40 \\ 15\overline{)600} \\ \underline{60} \\ 00 \\ \underline{00} \end{array}
\end{array}
$$

Instead of all those big numbers, you could just find $\dfrac{2}{3}$ of 60 and get the same answer.

60 times 2 and divide the answer by 3.

$$
\begin{array}{cc}
\begin{array}{r} 60 \\ \times\,2 \\ \hline 120 \end{array}
&
\begin{array}{r} 40 \\ 3\overline{)120} \end{array}
\end{array}
$$

using short division

Chapter Fourteen
In the Garden

Before we begin Chapter Fourteen, did you ever notice that four and fourteen have u's in them, but forty doesn't?

Spellling Enlish wurds is a lot harder than reducing fractions. $\dfrac{6}{8} = \dfrac{3}{4}$

Fred didn't know what to do. He had never seen a deer running loose on the KITTENS campus. There was a big sign in the rose garden that said . . .

Please do not pick the roses

Here are the nine thoughts that ran through Fred's head:

#1 The deer really is not picking the roses. He's *eating* them!

#2 The deer probably can't read.

#3 The deer probably can't even speak English, although I do know a horse that can.

#4 I wonder whatever happened to Miss Ente.

#5 If I tell the deer to stop eating the roses, it might just go on eating. I wouldn't know if he didn't understand me or if he was just pretending not to understand me.

#6 If I don't do something, instead of just two-eighths ($\frac{2}{8}$) of the rose patches being ruined, all the patches ($\frac{8}{8}$) will be wrecked.

#7 $\frac{8}{8}$ can be reduced to $\frac{1}{1}$ if I divide top and bottom by 8.

#8 100% means $\frac{100}{100}$

#9 $\frac{100}{100}$ reduces to $\frac{1}{1}$ if I divide top and bottom by 100.

Fred did a lot of thinking, and the deer continued to do a lot of eating.

Fred tried idea #5. He walked up to the deer and said, "Excuse me. I know that the sign says that you shouldn't pick the roses, but it probably also means that you shouldn't eat them."

The deer kept on munching the roses.

Fred had another four thoughts:

#10 Maybe I could build a big fence around the rose garden. Then the deer couldn't come and eat the roses.

#11 But if I build that fence right now, that would trap the deer inside the rose garden, and he could go on eating forever.

#12 Maybe I could tempt him with something better than roses.

#13 I have several pieces of food in my pockets. Maybe he would be interested.

Several pieces of food is an understatement. Fred had been sticking food in his pockets for years. He always said that the food was "for later," but he never got around to eating it.

Fred took a couple of pretzels out of his pocket and threw them to the deer.

The deer caught them in midair and ate them.

This might be working! Fred thought. If he is eating the stuff from my pockets, then he is not eating the roses.

Fred backed away from the rose garden and held out a piece of mushroom and pepperoni pizza.
The deer followed Fred and ate the pizza.

Digging deeper, Fred found a double cheeseburger. Fred started jogging toward the Math Building. The deer followed. Fred let him take a bite, then he ran a little farther, and then gave him another bite.

Cheeseburgers taste better than roses.

Now that the deer was away from the rose garden, Fred wished that the deer would stop following him.

At the bottom of one of his pockets, he found a basket of popcorn. Fred dribbled out a couple of kernels at a time and ran as fast as he could.

Fred got to the Math Building and ran up the two flights of stairs and down the hallway. He learned that deer can run at least as fast as people can. The deer followed him into his office.

The deer was in love with Fred.

Kingie was finishing up another oil painting. He shouted at Fred, "You didn't bring home another pet! We have had a cat and dogs and mice and fish. Enough!"

Fred could somehow sense that Kingie wasn't very happy with the idea of having a pet deer.*

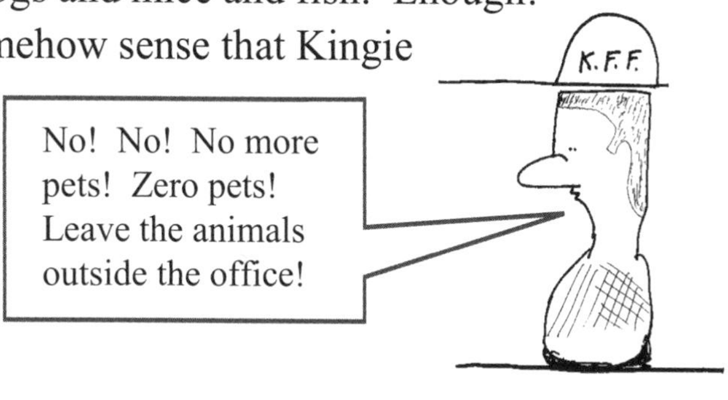

* This is called an understatement.

The deer found the desk drawers where Fred had stored a lot of his "for later" food. Fred opened the drawers, and the deer went to work devouring it all.

The deer burped. Fred didn't know that animals did that. He thought only humans burped.

When the deer couldn't find any more food in Fred's office, he wandered down the hallway and tried to get food out of the nine vending machines (four on one side and five on the other). When that didn't work, he headed down the two flights of stairs and was never seen again.

Moral: Sometimes deer (and people) only say that they love you when you have something to give them.

The evening edition of the campus newspaper hit the office door. Fred went and got it. (For obvious reasons, Kingie doesn't do a lot of walking.)

THE KITTEN Caboodle

The Official Campus Newspaper of KITTENS University   June 1 9:10 p.m.   10¢

exclusive interview

Campus Rose Garden Destroyed!

KANSAS: Ten minutes ago, Lawrence L. Wistrom, the head gardener at KITTENS, reported to the police that the world-famous rose gardens had been vandalized.

"Two of the eight rose patches were completely ruined," Larry said to the Caboodle news staff. "I have spent years creating that beautiful place on campus."

The police have no clue who did it since the two Coalbacks are currently in jail.   (more local news on page two)

Page two of the newspaper will be at the beginning of the next chapter. We need to take a little break with a *Your Turn to Play*. Please take out a sheet of paper and answer the questions.

Your Turn to Play

1. Fred had 13 ideas when he saw the deer eating the roses. If 5 of those ideas were good, what fraction of his ideas were good. (This is an easy question, but please write down your answer before turning the page. It will help the learning process.)

2. Two out of the 800 Caboodle readers (Fred and Kingie) knew who ate the roses. As a fraction that would be $\frac{2}{800}$

   Reduce this fraction.

3. Three-sixteenths ($\frac{3}{16}$) of the Caboodle readers cried when they read about the destruction of the rose gardens. How many cried?

   Do you need a hint? We are looking for $\frac{3}{16}$ of 800.

4. $40\% = \frac{?}{100}$

5. $\frac{7}{100} = ?\%$

. . . . COMPLETE SOLUTIONS . . . .

1. Five out of 13 can be written as $\frac{5}{13}$

2. $\frac{2}{800}$ can be reduced to $\frac{1}{400}$

if you divide top and bottom by 2.

3. $\frac{3}{16}$ of 800

       means 800 times 3 and divide the answer by 16.

$$
\begin{array}{r} 800 \\ \times\ \ 3 \\ \hline 2400 \end{array}
\qquad
\begin{array}{r} 150 \\ 16\overline{)2400} \\ \underline{16} \\ 80 \\ \underline{80} \end{array}
$$

    $\frac{3}{16}$ of 800 is equal to 150.

4. $40\% = \frac{40}{100}$

5. $\frac{7}{100} = 7\%$ You probably have never done one of these
               kinds of problems (going from a fraction to
               a percent).

               All the previous problems went from a
               percent to a fraction, as you did in
               problem 4.

There are 68 examples of how to do this on page 70. Just read
from right to left.

$$5\% = \frac{5}{100} \quad \text{becomes} \quad \frac{5}{100} = 5\%$$

Chapter Fifteen
Due by Midnight

F red turned to page two of the KITTEN CABOODLE.

page 2

University President declares

KITTENS University Needs a Seal

Anaheim, California: Seven minutes ago our president made an urgent phone call to the Caboodle newsroom.

Taking time out from his busy vacation schedule

"While standing in line waiting for Mr. Frog's Wild Ride," he said, "I noticed a lot of people wearing T-shirts from various universities.

"All the other universities each had a circle thing with some Latin words."

One of our reporters explained to the president that those are called the university seals.

"We need one!" the president shouted. "Start a contest or something and get one by the time I get back from my vacation." At this point he hung up.

Story continued on page three.

page 3

Our university needs a seal. Not this kind → Every other university has one. By law we can't print copies of the seals of other universities here in our newspaper. (Our lawyers told us that.) However, if you use your web browser in the images category and type in "university seals," you will see a zillion examples.

Submit your drawing by midnight and win big!

Fred searched on his computer and found the zillions of examples of university seals. It was now 9:15 p.m. He had less than three hours to work.*

Fred told Kingie that this was going to be one of the greatest pieces of art that he had ever done. Kingie had never

seen Fred do drawings except ones that looked like

"Ducky"

Fred told Kingie, "It is important to capture the real spirit of KITTENS University in its seal."

Kingie watched as Fred warmed up by drawing circles for his university seal.

---

* From 9:15 to 10 is 45 minutes. From 10 to midnight is 2 hours. Fred had two hours and 45 minutes in which to submit his entry.

Kingie went to his easel and painted »→

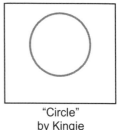

"Circle"
by Kingie

Kingie asked, "Is it okay if I work on designing the KITTENS seal also?"

Kingie received no answer. He looked and found Fred was fast asleep at his desk. Fred had had a long day. It was well past the time for five-year-olds to be in bed.

Kingie took Fred, tucked him into his sleeping bag, which was under his desk, and quietly wished him, "Goodnight." This was not the first time Kingie had put Fred to bed.

It was now 9:20. Kingie had two hours and forty minutes until midnight.

small essay

Fred's Art and Kingie's Art

When Fred wanted to do art, he was always in a hurry. If he had stayed awake, he would have drawn a big circle and then would have just filled it in with stuff that he knew how to draw.

This is called the sloppy approach to art.

Kingie did just the opposite. He knew that the drawing of the final submission was the *last thing* for him to do.

The *first step* is the most important. The first step is to think. All great art reflects the thoughts of the artist who has a deep vision of truth that needs to be conveyed. Great art—like great music and great literature—conveys a message.

We love a great artist, a great composer, or a great writer because we treasure the vision the artist has.

end of small essay

Kingie knew that designing a seal would not be great art. Instead, his challenge was to "capture the real spirit of KITTENS University"—as Fred expressed it.

He knew that just putting pictures of a mouse 🐭, a kitten 🐱, and a worm 🐛 inside a circle would not work. Ninety percent (90%) of Kingie's effort would be in finding the words and pictures that announce, "This is truly KITTENS University."

At first, Kingie thought of just putting a big picture of Fred in the middle of the seal. A lot of people, when they think of KITTENS, think of its most famous teacher, Fred Gauss.

But all the other teachers wouldn't be very happy with the university being represented by just one teacher—even if he was the best teacher.

In addition, Kingie worried what it would look like if this seal, which featured Fred, were submitted by Fred's doll.

Instead, Kingie knew that he needed to show the life of the students at KITTENS—what the students do during their waking hours.

Your Turn to Play

1. Ninety percent (90%) of Kingie's effort in creating the university seal is spent in thinking and planning. The remaining 10% of the time is used in making the final drawing.

    Change 90% to a fraction and then reduce the fraction.

2. The students are awake two-thirds ($\frac{2}{3}$) of every 24 hours. How many hours is that?

3. Over the years Kingie has seen how much time students at KITTENS spend in various activities. He noticed that the average student spends 25% of his waking hours going to class and doing homework.

    Change 25% to a fraction and reduce that fraction.

4. Using the results of questions 2 and 3, find out how many hours per day the average student spends doing student things (going to class and doing homework).

5. A class that meets one hour per week is worth one unit. If it meets three hours per week, it is worth three units. If a student takes 15 units per semester, he or she will usually graduate in four years.

    Many colleges charge students according to the number of units they are taking. That's called the tuition (two-ISH-shun).

    Suppose a university charged $79 per unit. How much tuition would a student taking 15 units be charged?

.... COMPLETE SOLUTIONS ....

1. $90\% = \dfrac{90}{100}$

To reduce the fraction $\dfrac{90}{100}$ we have to think of a number that divides evenly into both 90 and 100.

If we divide top and bottom by 10, $\dfrac{90}{100} = \dfrac{9}{10}$

2. $\dfrac{2}{3}$ of 24 means 24 multiplied by 2 and divided by 3.

$$
\begin{array}{r}
24 \\
\times\ 2 \\
\hline
48
\end{array}
\qquad
\begin{array}{r}
1\,6 \\
3\overline{)\,4^{1}8}
\end{array}
$$
Using short division

Students are awake 16 hours each day.

3. $25\% = \dfrac{25}{100}$

If we divide top and bottom by 5, $\dfrac{25}{100} = \dfrac{5}{20}$

If we divide top and bottom of $\dfrac{5}{20}$ by 5 again, $\dfrac{5}{20} = \dfrac{1}{4}$

(Or you could have divided top and bottom of $\dfrac{25}{100}$ by 25 and done everything in one step.)

4. We want $\dfrac{1}{4}$ of 16 hours. That means 16 times 1 and divided by 4.

$$
\begin{array}{r}
4 \\
4\overline{)\,16}
\end{array}
$$

Students spend 4 hours per day doing student things.

5. 15 units @ $79 per unit.
$$
\begin{array}{r}
79 \\
\times\ 15 \\
\hline
395 \\
79 \\
\hline
1185
\end{array}
$$

The tuition for 15 units would be $1,185.

Chapter Sixteen
Judging the Contest

Did you ever see a doll yawn? Some dolls never seem to get tired.* Their eyes remain wide open. It was 11 p.m. and Kingie had finished doing all the planning for the university seal he was creating.

He had one hour to do the painting and submit it. (Many artists would have drawn instead of painted. One artist who was a sculptor was carving his seal out of granite. There are many ways to do art.)

By 11:35 p.m. Kingie had finished his painting.

Alert readers may be wondering how Kingie was going to get his wet oil painting** to the KITTEN Caboodle office by midnight. Fred couldn't take it since he was asleep. Kingie couldn't walk that far. He couldn't call Betty or Alexander or any of Fred's other students since they were all asleep also.

The secret was hidden back in *Life of Fred: Goldfish*. In that book, Fred was in the restroom down the hallway from his office. He was using the world-famous KITTENS campus mail, which had a mailbox right next to the sink.

Kingie carefully wrapped his painting and carried it down the hallway to the mailbox.

KITTENS campus mail
Free delivery
Faster than email

---

* There are other dolls that love to sleep every night. When you tuck them in their beds, they never complain and head straight to sleep.

** Oil paintings take several weeks to fully dry.

With his very best handwriting, he had addressed the package.

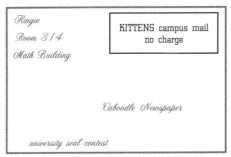

He dropped it in the mailbox and a special light turned on. A student letter carrier rushed over to the box, got the mail, and ran to deliver it.

Before Kingie got back to Fred's office, the package was already delivered to THE KITTEN CABOODLE.

Kingie went back to doing his regular oil paintings. He would work throughout the night while Fred was sleeping.

❀   ❀   ❀

Meanwhile, at the newspaper, contest entries were pouring in. The editor of the Caboodle told all of his staff that everyone would have to stay all night to judge all the entries.

When they announced the contest in the 9:10 p.m. edition of the newspaper and gave a deadline of midnight, they were

Caboodle Editor

hoping that there would be very few entries, and therefore they would have very little work to do. To their surprise, they were

receiving zillions of boxes, envelopes, cartons, containers, and packages.

The editor said, "It must have been because we wrote, "Submit your drawing by midnight and <u>win big</u>!""

One of the staff asked the editor, "What is the big prize?"

The editor shrugged his shoulders and said, "We haven't figured that out yet, but we have to determine the winner by the 6 a.m. edition of the paper. We have only about six hours. Let's get to work."

"Hey! Where do you want me to put this?" the campus mail carrier asked.

"Just stick it on the counter," he was told.

Just before midnight, the delivery truck rolled in with the largest delivery that had ever been made by campus mail. Before anyone on the newspaper staff could object, the truck slid the six-ton piece of granite onto the newsroom floor and left. No one could figure out how to move it. It would sit in the middle of the newspaper office for years to come.

Everyone on the staff thought that "Making brains like granite" was pretty dumb. Someone commented, "That means that we are all a bunch of blockheads."

Another asked, "Does that mean that KITTENS is in the stone age?"

When midnight came, the editor locked the doors so that no more entries could be submitted. There were 624 entries and 6 hours in which to find a winner. How many would they have to process each hour?

Ashley got an answer of **630** entries per hour by adding 624 + 6.

Chris got an answer of **618** entries per hour by subtracting 624 − 6.

Drew got an answer of **3,744** entries per hour by multiplying 624 × 6.

Joyce got an answer of **104** entries per hour by dividing 624 ÷ 6. (÷ means "divided by.")

Which answer was reasonable?

✖ If you processed 630 entries each hour, would you expect to process 624 in six hours? No.

✖ If you processed 618 entries each hour, would you expect to process 624 in six hours? No.

✖ If you processed 3,744 entries each hour, would you expect to process 624 in six hours? No.

If you wanted to process 624 entries in 6 hours, you would expect to do about a hundred each hour.

When you do a math problem, it really helps to look at your final answer and ask, "Is this reasonable?"

If you are asked to find the weight of a flower and your answer is 72 pounds, *it isn't reasonable.*

If you are asked to compute the distance between San Francisco and Detroit and your answer is 8 quarts, *it isn't reasonable.* Quarts is not even a distance.

If you are . . . Wait a minute! This is too much fun. It's now . . .

Your Turn to Play

1. Invent two more "If you are asked . . ." statements that have silly answers. They are fun to make up.

2. Six tons equals how many pounds? (One ton = 2,000 lbs.)

3. Three-fourths ($\frac{3}{4}$) of the 624 entries were assigned to Joyce to look at. How many was that?

4. Change 75% into a fraction and then reduce the fraction.

5. Is this a function?

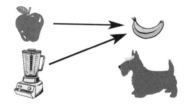

Domain   **Codomain**

6. Is it possible to create a function whose domain is the natural numbers $\{1, 2, 3, 4, 5, 6, \ldots\}$ and whose codomain is the state of Kansas?

.... COMPLETE SOLUTIONS ....

1. Your answers will probably be different than mine. Here are some of my favorites.

If you are asked to find the number of calories in a slice of cheesecake and your answer is 2 calories, *it isn't reasonable.*

If you are asked to find the price of the university president's mansion and your answer is 13¢, *it isn't reasonable.*

If you are asked to find the area of Fred's desktop and your answer is 29 square miles, *it isn't reasonable.*

If you are asked to find the time it takes to eat dinner and your answer is 832,683,552 years, *it isn't reasonable.*

If you are asked to find the population of Reno, Nevada and your answer is 705, *it isn't reasonable.*

2. Six tons equals how many pounds? (One ton = 2,000 lbs.) The General Rule is if you don't know whether to add, subtract, multiply, or divide, restate the problem using easier numbers and notice which of the four operations you used. For example, suppose we had 3 tons and each ton was equal to 2 pounds. Then 3 tons would be equal to 6 pounds. We multiplied.

$$\begin{array}{r} 2000 \\ \times \ \ 6 \\ \hline 12000 \end{array}$$ Six tons equals 12,000 pounds.

3. $\frac{3}{4}$ of 624 means 624 times 3 and divide the result by 4.

$$\begin{array}{r} 624 \\ \times \ \ 3 \\ \hline 1872 \end{array}$$ $4\overline{)1\,8^27^32}$ using short division

$\frac{3}{4}$ of 624 is 468.

4. $75\% = \frac{75}{100} = \frac{3}{4}$ (after dividing top and bottom by 25)

5. Yes. It is a function. Each member of the domain has exactly one image in the codomain.

6. It is possible. In fact, there is only one possible function. It is the rule: *Assign each natural number to Kansas.*

Chapter Seventeen
Many Entries

The contest began at 9:10 p.m. and ended at midnight. That is less than three hours. No one on the KITTEN CABOODLE news staff could figure out how 624 entries could have been submitted.

Additionally, most of the students and teachers were on vacation since the university president had declared a nine-day holiday before graduation.

When Ashley, Chris, Drew, and Joyce started opening the boxes, envelopes, cartons, containers, and packages, the mystery was solved.

When Ashley opened his first envelope, he found a scribbled drawing:

He said, "It must have taken someone three seconds to make that drawing. How could they have hoped to win?"

When Chris opened a carton, he found an identical entry:

Joyce quickly opened a half dozen envelopes and found:

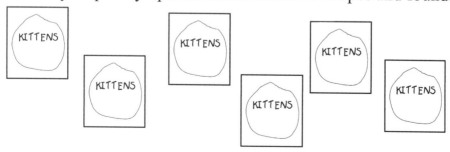

Drew checked who had submitted all these identical entries. They were all from the same person: Joe.

Joe was the student who liked jelly-bean sandwiches.

The phone rang and Joyce picked it up.

"Hi. This is Joe. I'm just calling to find out if I won the drawing."

Joyce explained to Joe, "This event was not a raffle in which the winner is picked at random among the entries. This was a contest. The entries are judged. We are trying to select the best seal for KITTENS University."

Joe said, "Oh," and hung up.

Ashley, Chris, Drew, and Joyce opened the entries as fast as they could. It was not going to be a long night of judging 624 different entries.

Six hundred and twenty-two entries were all alike . . .

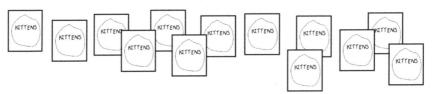

and they were all thrown into the trash can.

$$\begin{array}{r} 624 \\ -\ 622 \\ \hline 2 \end{array}$$

That left two entries that they would have to choose between. They wouldn't have to work all night. They just had to pick which one of the two was better, and they could go home and go to sleep. The choice was between the stupid 6-ton rock and a wet oil painting that was sitting on the counter.

The four judges were frightened.

Chris said, "What if that painting is as ugly as Joe's entries? Then we would have to choose the dumb rock."

Drew said, "Nothing could be as bad as Joe's entries." Drew was *whistling in the dark.**

All four were saying, "Please, please, make this be a good one."

And it was. It was gorgeous.

---

* *Whistling in the dark* is an idiom. It means to try to remain brave when facing an unknown danger. It is an attempt to talk yourself into courage.

Ashley said, "It even looks like a university seal should look like."

Chris said, "It gives a bit of our history. Our first classes began in 1929.*"

Drew said, "It captures the spirit of KITTENS by listing the four things our students do: study, play, pizza, and nap."

Joyce said, "Hey! It's even got some Latin. That adds real dignity."

Ubi est mea anaticula cumminosa?

Ashley said, "It is some kind of question."

Chris asked, "How can you tell? You don't know Latin, do you?"

Ashley said, "No, I don't, but there is a question mark at the end."

Drew asked, "But what does it mean? All I know is that *est* means 'is' and *mea* means 'my'."

Joyce suggested, "Maybe it means something like *What is my calling in life?* That would be a great motto for a university."

small essay
A Great University

Universities should exist for one purpose: *To change students.* If you spend four years of your life and tens of

---

* For those who know a little history, 1929 wasn't an especially good year for new beginnings. That was the year in which the stock market crashed and the Great Depression began.

thousands of dollars attending a university, and all you have to show for it is:

✍ memories of football games

✍ a head full of facts such as "K stands for potassium"

✍ a big student loan

✍ a paper degree

then that university failed in its mission.

If you want memories of football games, watch them on television. If you want debt, buy a fancy car. If you want a head full of facts, watch quiz shows on television. If you just need a degree, they are easy to get on the Internet.

Good universities will help you get a decent job.

Great universities help you understand who you are, what your place is in society, and what your place is in the universe.

Do you see the difference between *good* and *great*?

end of small essay

Your Turn to Play

1. Teaching a university course in history would be a perfect place to help students learn about their place in the story of mankind. Instead, in most history classes, the teachers take the easy road and just have the students memorize dates and names. That is why many history classes are b-o-r-i-n-g.

According to one survey, 80% of history classes just teach facts and not meaning.

If there are 795 history classes being taught in your state, how many of them just teach facts?

.... **COMPLETE SOLUTION** ....

1.   We want 80% of 795.

      That means $\frac{80}{100}$ of 795.

To make our work easier, reduce the fraction $\frac{80}{100}$

What number divides evenly into both 80 and 100? Ten.
Divide top and bottom by 10.

$$\frac{80}{100} = \frac{8}{10}$$

Can $\frac{8}{10}$ be reduced any further? Yes. Both 8 and 10 are even,
so we can divide top and bottom by 2.

$$\frac{8}{10} = \frac{4}{5}$$

     We want $\frac{4}{5}$ of 795.

     That means 795 times 4 and divide the result by 5.

$$\begin{array}{r} 795 \\ \times\ 4 \\ \hline 3180 \end{array} \qquad \begin{array}{r} 636 \\ \hline 5)\overline{3180} \end{array}$$

       636 of the 795 history classes just teach facts.

Note: You could have reduced $\frac{80}{100}$ down to $\frac{4}{5}$ in one
step if you had divided top and bottom by 20. Most people
would not have noticed that.

Chapter Eighteen
The University Motto

The four staff members of THE KITTEN CABOODLE were excited. They didn't know that their university motto was Ubi est mea anaticula cumminosa? They decided that this should go right below the masthead of their newspaper.

THE KITTEN Caboodle
Ubi est mea anaticula cumminosa?

The Official Campus Newspaper of KITTENS University

Ashley: What dignity!
Chris: It makes us look smart!
Drew: And it's in Latin!
Joyce: But what does it mean?

None of the four realized that Ubi est mea anaticula cumminosa? wasn't the real motto of their university. It was just made up by Kingie when he designed the university seal. Kingie knows more Latin than most people.

Ashley: I'm kinda curious what Ubi est mea anaticula cumminosa? means. Do any of you have a Latin dictionary?
Chris: No.
Drew: No.
Joyce: No, but I'll check to see if the editor has one.

Joyce came back with the dictionary. He said, "The word *ubi* means 'where.'"

Chris: So Ubi est mea means "Where is my. . . ." So Ubi est mea anaticula cumminosa? probably means something like "Where is my real goal in life?"

Drew: That sounds right.

Joyce: When I look up cumminosa all I find is cumminosus. Latin has lots of different endings on words. The definition of cumminosus is "full of gum" or "gummy."

Ashley: Weird. But you've got to remember that Latin was used hundreds of years ago. You won't find modern words like *picnic*, *radio*, or *thermonuclear* in traditional Latin. The word cumminosa might mean anything chewy or gluey or bouncy.

Chris: Well, I guess a lot depends on what anaticula means.

Everyone held their breath as Joyce looked up anaticula.

Joyce: This doesn't make sense. The word anaticula means a baby duck. This is crazy. Ubi est mea anaticula cumminosa? can't mean "Where is my gummy baby duck?"

Then Chris started laughing. And laughing. And laughing. His face turned red. Tears rolled down his cheeks. He could hardly speak. Everyone looked at him.

Chris finally said, "Don't you get it? (giggle) It's not 'Where is my gummy baby duck?' (laugh) Ubi est mea anaticula cumminosa? means 'Where is my rubber ducky?'"

And that is how KITTENS University acquired its official motto. A five-year-old doll named Kingie was the inventor. This seal would be placed on all the official correspondence from the university and on every diploma. The university president would quote the motto in his speeches but never learn what it meant. It just sounded good. Parents loved the seal with the Latin words on it. It meant that they hadn't wasted their money on their kids' educations.*

❀   ❀   ❀

The four staff members presented the painting to the editor. The editor looked at it for a moment and said, "That looks like a good choice. Everyone's going to like that bit of Latin, whatever it means."

No one said a word.

The editor asked, "Who did that beautiful painting?

"All we know," said Chris, "is that it is someone named Kingie who has an office here on campus."

"I never heard of him," the editor said. "That's even better. When you write up the article for the 6 a.m. edition that announces the winner, be sure to play up the fact that the winner is someone who is unknown. People like to root for the little guy."

Ashley was assigned to find a picture of Kingie for the article. Since it was 1 a.m., he couldn't go to room 314 and wake him up to get a photograph. Instead, Ashley would have to look through the files to see if there were any existing photos.

---

* Plural words that end in *s* just get an apostrophe.
All other words get an apostrophe and an *s*.

Plural words ending in *s*	All other words
tourists' bags	Jane's hat
Cub Scouts' lunches	Venus's beauty
kids' education	media's bias

Media is a plural word that does not end in *s*. *Medium* is its singular. Books are a medium. Television is a medium. Radio, books, and television are some media.

Chris was assigned to find background information about this guy named Kingie. Was he a teacher? Was he a custodian? When did he come to KITTENS? How old is he?

Drew was assigned to look into the Caboodle archives to see if Kingie had appeared in any newspaper articles in the past.

Joyce was to take the material that Ashley, Chris, and Drew found and write up the article.

Ashley found no photos. Chris couldn't find anyone named Kingie in the faculty list or in the staff list. Drew looked through the newspaper files and found that he had never been mentioned.

Joyce said, "If it weren't so early in the morning, we could just send a reporter and a photographer to room 314. We could interview this guy and take some pictures. The article would be easy to write."

Joyce didn't realize that the light in room 314 was on. Kingie was wide awake doing his oil painting. How many times in life does each of us experience: *If only I had known.*

small essay

If Only I Had Known

If this were a traditional math book, it would only tell you math facts. For example, it would say If you want to average 4 numbers you add them up and divide by 4. The average of 6, 8, 5, and 9 is (6 + 8 + 5 + 9) ÷ 4 = 28 ÷ 4 = 7.

However, you may have noticed that this is *not* like any other math book. In the Life of Fred™ books, we offer a more complete education, which includes history, English, biology, physics, economics, poetry, foreign languages (*Ente* = duck in German, *café* = coffee in French), and what it means to be a great university. We now offer . . .

A List of Some of the Things
That I Wish I Had Known
Earlier in My Life

1. After about the age of 25 or so, your metabolism starts to slow down. If you eat the same stuff that you ate when you were a teenager, you will gain weight.

2. If you don't live alone, there are two rules you have to follow.

Rule #1: If you fill it up, then empty it. (Like a trash basket.)

Rule #2: If you empty it, then refill it. (Like a toilet paper dispenser.)

3. The most important thing in life is not food. Is not money. Is not fame. Is not math. Is not clothes. Is not health. Is not your country. Is not a long life. Is not having a fun time.

4. The second most important thing in life is discovering what the most important thing in life is.

end of small essay

Your Turn to Play

1. Since this is also a math book, find the average of these 7 numbers: 9, 6, 3, 4, 9, 90, 12.

2. Is it possible to create a function in which the domain is {5} and the codomain is {A, B, C}?

3. Is it possible to create a function in which the domain is {7, 8, 9} and the codomain is {✿}?

4. Find 20% of 445.

·······COMPLETE SOLUTIONS·······

1. To find the average of 9, 6, 3, 4, 9, 90, 12, you add them up and divide by 7.

$$
\begin{array}{r}
9 \\
6 \\
3 \\
4 \\
9 \\
90 \\
+\ 12 \\
\hline
133
\end{array}
$$

$$7 \overline{)1\,3\,^6 3} \qquad \text{using short division}$$

The average of 9, 6, 3, 4, 9, 90, 12 is 19.

2. There are several possible functions in which the domain is {5} and the codomain is {A, B, C}.

    Example #1: 5 → A Then every element of the domain has exactly one image in the codomain.

    Example #2: 5 → B Then every element of the domain has exactly one image in the codomain.

There is a third example, but I won't mention it.

3. There is only one function in which the domain is {7, 8, 9} and the codomain is {✿}. That is the function given by the rule: *Map 7, 8, and 9 to* ✿. Then every member of the domain has exactly one image in the codomain.

4.   20% of 445

$$\frac{20}{100} \text{ of } 445$$

Reduce the fraction $\frac{20}{100}$ by dividing top and bottom by 10 and then by 2: $\frac{20}{100} = \frac{2}{10} = \frac{1}{5}$ (or you could do it in one step by dividing by 20)

$\frac{1}{5}$ of 445 means 445 times 1 and divide the result by 5.

$$
\begin{array}{r}
445 \\
\times\ \ 1 \\
\hline
445
\end{array}
$$

$$5 \overline{)4\,4\,^45}$$
$$\ 8\ 9$$

20% of 445 is 89.

Chapter Nineteen
A New Morning

F red awoke. It was 5 a.m. It was summer and the sky was starting to get light. Kingie had been painting since midnight and had just finished his fifth oil painting.

Fred put on his jogging clothes to get ready for his morning jog. He looked at his desk and noticed the circles he had drawn before he fell asleep.

 "Oh no," he said to Kingie. "Did I fall asleep? Have I missed the deadline for the university seal contest? Did you put me to bed last night?

Kingie was cleaning his brushes and answered, "Yes. Yes. Yes."

Fred felt bad. He hadn't meant to fall asleep, but he had. Even if he had entered the contest, he would not have won.

First Place   Second Place   Third Place   Last Place

But jogging often made Fred feel better. He headed down the hallway past the nine vending machines (4 + 5), down two flights of stairs, and out into the morning air.

One of the good things about jogging was that it gave him time to think. As he ran past the campus tennis courts and the university chapel, and down Archimedes Lane, he was grateful:

⋆ that he had Kingie who would put him to bed when he fell asleep at his desk,

⋆ that he was healthy and could go jogging,

✫ that he knew a lot of wonderful mathematics,

✫ that he could teach math every day that school was in session,

✫ that he had a lot of students who loved him. . . .

The list of things that Fred was grateful for went on and on. The fact that he had missed a chance to enter the university seal contest r e a l l y   w a s n ' t   t h a t   i m p o r t a n t .

At the first moment of sunrise, Fred started singing. He made up the song as he ran. As he passed the university rose garden, he sang, "Oh brother roses, ♪♬ Oh sister birds. ♬ Come sing with me."

The tune wasn't very good. Fred's singing wasn't very good. But that r e a l l y   w a s n ' t   t h a t   i m p o r t a n t .

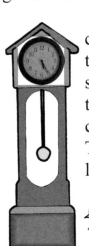

Fred ran past the university grandfather clock. It is one of the largest grandfather clocks in the world. The university president had ordered it several years ago. He said he wanted a twenty-foot-tall clock. When it arrived, he realized that the ceilings in his mansion were only eighteen feet. The clock was put outside. In math, "eighteen is less than twenty" is written as $18 < 20$.

As Fred jogged by the clock he noticed that he was about 15% as tall as the 20-foot clock.

He did the math in his head: 15% of 20

$$\frac{15}{100} \text{ of } 20$$

Dividing top and bottom by 5 $\frac{3}{20}$ of 20

That means 20 multiplied by 3 and divided by 20

$$\begin{array}{r} 20 \\ \times\ 3 \\ \hline 60 \end{array} \qquad 20\overline{)60}^{\,3} \qquad$$ 15% of 20 is 3.

Fred is three feet tall.

Fred counted by fives. It was 5:25 or 35 minutes until 6. Fred wanted to get back to his office by 6 to read the morning edition of THE KITTEN CABOODLE.

Fred took one loop around the Great Lake that was near the campus and came back to his office at exactly 6 a.m.

The morning paper arrived just as he opened his office door.

The paperboy shouted, "Sorry, sir. I didn't see you."

Fred responded, "It's quite all right. No harm done."

He opened the newspaper and read the headline.

THE KITTEN Caboodle

The Official Campus Newspaper of KITTENS University   June 2   6 a.m.   10¢

just announced

Kingie Wins University Seal Contest

KANSAS: The judges all agreed. Kingie's oil painting has won first place in the contest.

Copies of our new university seal are being printed up and will be posted everywhere around the campus.

The only thing known about the winning artist is that his return address was Room 314, Math Building.

Fred handed the paper to Kingie and said, "This is wonderful. Congratulations."

There was a knock on the door.

Kingie packed up his easel, brushes, and paints, rushed into his fort in a corner of Fred's office, and shut the door. (This was the one-foot-tall fort that Kingie had made when Fred had brought home a cat as a pet.)

> Tell them I'm not available!

Four reporters, a photographer, and hundreds of people who had come to find out who Kingie was were outside of Fred's office.

Kingie was shy. Many people who are six inches tall don't like to meet hundreds of people. Kingie didn't want fame. He just wanted to do art.

Fred, on the other hand, is three feet tall. He invited everyone into his office. Only 40 people at a time could fit. He chatted with them and told them all about his doll. When they left, another group of 40 came in, and Fred answered their questions.

From 6 a.m. to 1 p.m. (7 hours) Fred entertained the crowd.

During that time Kingie had painted four more paintings. Being famous was not important to him—doing art was.

Index